Six Inches _{of} SOIL

Praise for *Six Inches of Soil: How to Heal Our Soils, Ourselves, and Our Communities Through Regenerative Farming*

'*Six Inches of Soil* tells the story of how and why agriculture went down the path of soil degradation which has led to losses in biodiversity, carbon, water quality and nutrient density in our foods BUT it then takes you on a captivating journey of regeneration. Highlighting farms on their regenerative journey. Showing us the way and giving us hope. A must read for all society.'

GABE BROWN
Rancher, Educator, Author (*Dirt to Soil*) and Student of the Soil

'Readable, poetic and illustrative in its production; charming, evocative and educational in its content, *Six Inches of Soil: How to Heal Our Soils, Ourselves, and Our Communities Through Regenerative Farming* is above all helpful and honest. It's not just for farmers, but for anyone and everyone.'

JENNY JEFFERIES
author of *Islands In A Common Sea: Stories of Farming, Fishing and Food Around The World*

'The team behind *Six Inches of Soil* have pulled off something remarkable. This book has succeeded in capturing the passion and wonder from the film, supplementing that content with scientific, farming and policy depth to produce a work that stands strongly on its own feet. If you are lucky enough to have already watched the film, you will be delighted by the extra detail and discourse within these pages. If you are reading it fresh, the combination of interviews, discussion and reflection will probably change the way you look at soil forever.'

BEN RASKIN
Head of Horticulture and Agroforestry, Soil Association
& author of *The Woodchip Handbook*

'To feed a growing global population, we will need to produce more food than ever before – but we must do so more sustainably than has been the case in the past seventy years. *Six Inches of Soil* provides a hopeful vision for sustainable farming, based on careful stewardship of our most precious resource, alongside farming with – rather than in opposition to – our natural environment. A timely work which points the way to a more sustainable approach to food, farming and the environment.'

JOE STANELY
Head of Sustainable Farming, The Allerton Project & author of
Farm to Fork: The Challenge of Sustainable Farming in 21st Century Britain

'A fantastic assembly of the trials, tribulations, successes and failures of all those connected to the land that are looking at replacing death with life in the food production system.'

BEN TAYLOR-DAVIES/REGENBEN

SIX INCHES OF SOIL

How to Heal Our Soils, Ourselves, and Our Communities through Regenerative Farming

Edited by
Molly Foster, Priya Kalia & Jeremy Toynbee

First published 2024
Copyright © Six Inches of Soil Ltd

Published by
Unbreaking
An imprint of 5m Books Ltd
Lings, Great Easton
Essex CM6 2HH, UK
Tel: +44 (0)330 1333 580
www.5mbooks.com

Follow us on
Twitter @5m_Books
Instagram 5m_books
Facebook @5mBooks
LinkedIn @5mbooks

A Catalogue record for this book is available from the British Library

ISBN 9781917159005
eISBN 9781917159043
DOI 10.52517/9781917159043

Book design and layout by Alex Lazarou
Six Inches of Soil title graphic by Brook Morgan
Line art, page footers, Appendix II, www.vecteezy.com/free-vector/farm, Farm Vectors by Vecteezy
Photographs by and © Six Inches of Soil Ltd, unless indicated otherwise
Printed by Hobbs the Printers

Contents

Foreword

SARAH LANGFORD

I once watched soil move a woman to tears. She had been part of a group of people I was talking to at a festival. Polite and interested but largely urban, my audience had listened patiently as I explained how changing the way we looked after land could restore soils ravaged by many decades of intensive farming. They watched with curiosity as I set two large jars of water upon a table in front of them and balanced two clods of earth in a scoop of chicken wire in the top portion of the jars. One lump of suspended soil began to bleed into the water, turning it to silted brown in seconds. The other lump held its shape, the water staying crystal clear. This was, I said, the slake test. It was a way of showing what happened to the structure of soil when you tilled it and left it bare of plants for long periods. The disintegrating soil had been taken from a conventional arable field; the second I had dug up from neighbouring pasture.

The day was hot and beautiful. On a whim, I suggested we go outside into the pasture field. We stood in the sun as I dug another, deeper hole. On my knees, I flipped over the sward lid so that they could all see and smell the earth underneath. 'Touch it!' I encouraged them. 'Feel how cool it is, even in the heat of the day.'

The woman was the last to take me up on the invitation. As we knelt opposite one another I watched her sink both hands into the earth and, as she brought a fistful up to her face, breathing in its smell, her eyes filled with tears. She came up to me afterwards and held my arm. 'That is the first time I have ever felt the soil', she said. 'I live on a farm! But I've never touched its soil.' She said she had called her husband afterwards and told him of her revelation and of the farming movement that was regenerating both land and our connection to it. 'Yes!' He had said to her. 'Yes! We are doing all this; I've been trying to tell you…'

I begin the foreword with this story to demonstrate three things. Restoring our soils can change our landscapes: increasing biodiversity above and below ground, improving land's water holding capacity, sequestering carbon and cleaning air. Restoring our soils can change the nutritional value of the food we eat, enabling natural cycles to deliver complex micro-nutrients which are not available from a bag. But restoring our soils can also change our relationship with, and connection to, the earth. Attention is, after all, the beginning of devotion.

Within this book you will find advice and explanations about our food and farming system and the science of soil from experts who have been in this field (often literally) for many years. But this is a book which goes beyond a practical guide or academic analysis. The writers, farmers, growers and commentators in these pages prove that the world beneath our feet holds both wonder, and answers.

The human stories that are woven around the factual analyses in the pages you are about to read do more than show us how a change in the way we produce food and manage landscapes can create a resilient system, both biologically and economically. They show us that farming with nature, not against it, changes people. It transforms not just the land but the way those who care for it feel about what they do, and who they are. In restoring their soils and increasing the number of species they share their land with, those in this book have found their own connection with the land, and one another, transformed too.

There is much in the wider world which might leave us in despair. The unintended costs of modern, industrial farming spelled out in some of the following pages are bleak. But as with the film, which accompanies this book, there is far more hope than helplessness. Change is coming. It has to. It takes, as it always has, the bravery and courage of those who try something different to those around them to withstand the heat of scepticism. It is hard being the outlier when people are telling you it cannot be done. The reason this book is so important is not just that it shows this way of growing food is possible. It shows others who want to do the same that they are not alone. It shows people trying to make good decisions in a world whose problems can often feel too big to influence that food choices are our greatest point of agency. And it shows too that in a world where mystery and awe often feel absent, sometimes the simple act of putting our hands in the earth can become a moment that moves us to tears. The answer to all of it lies right beneath our feet, in just six inches of soil.

"Despite all our accomplishments,
we owe our existence to a
six-inch layer of topsoil and
the fact it rains."

PAUL HARVEY (1978)
US radio broadcaster

Introduction

CLAIRE MACKENZIE, COLIN RAMSAY & LUCY MICHAELS

Long live the earthworm!
Long live the farmer!

These were the words of the wonderful Satish Kumar as he closed the UK première of *Six Inches of Soil* at the Oxford Real Farming Conference in January 2024. The energy in the room was palpable and we were delighted to have received such a warm reception to the film. Satish commented that he'd never seen a film in which earthworms came on screen so often. While worms are obviously the real stars of the show, the stars of the evening and of this story are the three new entrant farmers who have bravely shared their stories with us.

Industrial farming has transformed Britain's rural landscapes. The post-Second World War shift towards more and more intensive farming significantly increased crop yields, helping to reduce hunger and kick start the economy. Yet this has all come at a terrible cost. Industrial agriculture has taken a huge toll on biodiversity, polluted our seas and freshwater sources, turned animals into little more than factory-produced products, depleted our soils of nutrients and emptied rural areas of meaningful sustainable jobs. Industrial agriculture also makes a significant contribution to climate change and is coming under increasing pressure to change its ways.

A handful of supermarket retailers and food processing companies now control food production in Britain, with nine retailers making 94.5% of food sales in Britain. As a society, we've become so disconnected from the ways in which our food is produced, packaged and transported. Most of us seem happy with the 'choice', 'convenience' and 'good value' that supermarkets seem to offer, but we are also addicted to ultra-processed food in a way that is contributing to an unfolding public health crisis.

But **change is in the soil** – pioneered by a quiet but rapidly growing food and farming movement in the UK that seeks to completely overturn the way we have farmed and eaten over the last 70 years.

We began developing the film in early 2021 and it has been a real labour of love. Our journey started with the making of a short film, *From the Ground Up,* in 2020 for South Cambridgeshire Council and the charity Cambridge Carbon Neutral, which was a key inspiration for us. We've told many times the story of standing in a field of 7 ft (2.1 m) sunflowers in the middle of winter and looking over at a neighbouring conventional farm where the soil was like house bricks. It was a massive penny-drop moment for us. We felt compelled to make the film because we were so inspired. Inspired by seeing what's possible on a farm, by the amazing people who grow our food in a nature-friendly way and by an alternative vision of the food system.

Propelled by crowdfunding, generous private donations and countless voluntary hours, we have been able to shoot, edit and promote the film as a completely independent feature-length documentary. It's been a whirlwind, and we're so grateful to everyone who has been on this journey with us. And now, in this book, we can take the time to dig a little deeper: deeper into the science of some of the problems and solutions in food and farming; deeper into the stories of the new entrant farmers in our documentary; and deeper into the experience and opinions of the experts featured.

Six Inches of Soil has truly come together through relationships. We began as a small team nurturing the seed of an idea, with no funds and this huge challenge of depicting the complexities of the soil and our food system in a film. But we just knew this story had to be told.

Drawing on our different strengths and connections, the team reached out to anyone who could help us. And it was those relationships that helped us build the story, an audience and a community around this film – in a truly mycelial way. As our team has grown, we have nurtured, inspired and fed each other. And we have to thank the living soils for that inspiration. We also think that this relational way of working is part of why people have been so touched by the film. The stories of our farmers are unquestionably moving, but the film also asks bigger questions that hint at far deeper human challenges that many of us feel but can't quite articulate.

How has it come to this point in our history that we hardly value the food we eat and the soil that it's grown in, and care little how food production impacts animals and the environment? What does it mean to be now, for many of us, among the second, third, fourth generations that are urbanised and no longer have a close relationship to the soil or the farmscape? A relationship that has been integral to human life since the beginning of agriculture some 12,000 years ago.

By asking these questions, truly transformational change can happen. Agroecology, building relationships with where our food comes from and putting our hands back in the soil, can be a more fulfilling way for us, as humans, to exist in the world.

Through *Six Inches of Soil*, we want to give a platform to this movement as it grows in Britain. It is a story of courage, vision and hope. We are showing why a new generation of farmers is turning away from conventional farming and choosing to work with nature to create resilient farming systems that do not rely on chemical inputs, heavy mechanisation and monocrops. Through these farmers' eyes, we see the highs and lows of changing a 'broken' food system. How they're healing the soil and the water, boosting biodiversity and fighting climate change while providing healthier, more nutritious food.

We feature three young, new entrant farmers: Anna Jackson, an 11th-generation farmer, and her dad Andrew on a mixed farm in north Lincolnshire; Adrienne Gordon a market gardener near Cambridge; and Ben Thomas a cattle farmer in Cornwall, who specialises in pasture-fed beef. Each of these stories gives a different take on how principles of regeneration can be applied to different production systems and on different scales. There are particular barriers for new entrant farmers in the UK. Join them as they tackle the trials and tribulations of starting a new business and see how they break into an industry that is famously hard to establish a foothold in. We follow them as they navigate change and get to know their land. Each of them visits and receives advice from more experienced mentors showing them what's possible when we pay more attention to nature and to relationships in our food production.

By highlighting the amazing work of brave new entrants and experienced pioneers we hope to inspire many other farmers to start their journey – a journey of reducing artificial inputs, reducing tillage, working in harmony with nature and supporting a healthier soil system. But this book is *not just for farmers*. Reconnecting with our food, and regenerating our soils, ourselves and our communities benefits everyone and needs everyone to be involved. We want to inspire farmers with the confidence and practical know-how to adopt regenerative farming approaches. We want to give people the impetus and information to rethink their food choices. And we hope that contributing to a groundswell of public opinion can lead to changes in policy, support and funding for a British regenerative farming and an agroecological revolution. We want to challenge the prevailing idea of food scarcity and present practical pathways towards ensuring nourishing and genuinely sustainable food.

We hope that our audience and readers can take a moment to reflect on these important questions, reflect on our values and commit to rediscovering our regenerative human natures. If we can, then this film, this book and this moment we have opened up can truly lead to lasting change.

What's in the book?

Throughout the process of making the film we were lucky to interview numerous experts from the world of food and farming while also following the progress of our new entrants. In this book we can now share with you more of what we gathered that just wouldn't fit into a 90 minute film. We begin with being guided, by the words of our experts, through the major challenges of today's food system, exploring why it's been called 'both a miracle and a disaster'. In Chapter 2, we look more closely at what we mean by regenerative farming. We are led here by Marina O'Connell in an exploration of the different movements of nature-friendly farming over time, through biodynamic, organic, permaculture, agroforestry, agroecology and regenerative farming. This is an area in which there is often confusion and sometimes a lack of clear definitions but also one where we can see fundamental connections in the desire to resist industrial methods and embrace ecological alternatives. Chapter 3 explores in more detail the soil science that informs the present-day movement and discusses questions around carbon sequestration and measuring. By starting here, we hope to give you a firm rooting in why we think these farming practices are so important, before we go on to see how they're put into action.

The midsection of the book (Chapters 4–6) offers a deeper dive into the stories of the three entrant farmers from the documentary and explores in more detail the experiences and wisdom of the mentors. The land Anna, Adrienne and Ben care for is diverse; different in area, typography, soil type and rainfall. Their approaches are correspondingly different. Nonetheless, they have much in common. They share a desire at their very cores to be better custodians of their land and animals, and to bring about change. Each uses multiple regenerative practices and between them they demonstrate the full range. Sadly, but most significantly, they also share the challenge of earning a fair living from farming and growing within the current poorly structured agri-food system.

Access to and transfer of land are fundamental to all their stories. In Chapter 4, Anna and her dad, Andrew, are self-admittedly in a stronger position than Adrienne and Ben who are tenants. They own their land and it gives them greater freedom to experiment. The question of inter-generational transfer of control is fascinatedly woven through their story. Anna having returned to the family farm has to find her place in the system and to negotiate with her dad the rate of progress towards and the nature of their goals. They share a similar overarching approach and motivation but there's variation in priorities. As Anna learns and becomes more confident, Andrew has to adjust and cede control. On many farms this process can be fraught with tension and disagreement; Anna and Andrew show a way forward based on mutual respect for each other and a good dose of humour.

Adrienne (Chapter 5) and Ben (Chapter 6) both took land-access opportunities that were only available because of the progressive, regen-erative landlords who own the land they work. Adrienne's experience will be recognisable to many horticulturalists and small-scale growers: years spent volunteering and travelling around; all the time looking for a chance, looking for a plot of land. The serendipitous reading of a notice-board post by her landlord, Tom, led to the invitation to return to her home county, Cambridgeshire, and a lease on 1.6 ha. Tom added some basic infrastructure (fencing and rainwater collection) and waived rent initially. Why? Because he is on his own regenerative journey on his arable farm and wanted to broaden the farm's offering, to connect and to build community.

Ben's land has a long history of being leased out. An intergenerational change in the ownership led to a change in approach. A regenerative custodian was sought to restore the soil, and Ben answered the call. A partnership was struck with Ben providing the labour and skills and they the land and buildings. Costs and profit initially were shared within an agreement that sees the share move favourably towards Ben over time.

Their three stories are inspiring, guiding and frustrating. *Six Inches of Soil* was conceived to bring such stories to greater attention: through them to inspire others to follow their example, to offer guidance by showing what is possible and to engender frustration at the challenges they have to overcome. Please allow yourself to be inspired, to be guided and to turn your frustration in to action.

We then take a step back from their individual stories, thinking about social regeneration, communities and the spiritual components of agroecology in Chapter 7, before rounding off with our vision of the future of the agri-food system in the UK (Chapter 8). A thread that will follow us throughout these stories – from the most scientific to the most personal – is that of relationships. While this will really come to the fore in Chapter 7, throughout the book we cannot get away from the importance of relationships – relationships in ecosystems from the smallest soil microbes to whole-farm level, human relationships to the land and to the rest of nature, and relationships in human communities.

Interspersed between the substantive chapters you'll find six concise interludes. Four that address some key issues (land-use policy, greenwashing, subsidies and food security) that are touched on throughout and two that explain, with case studies, practices fundamental to the journeys of our farmers (agroforestry and enterprise stacking).

The epilogue is our call to action. Challenging you to act now.

At the back of the book you will find appendices introducing our partner organisations and providing more detail on the advising farmers. Brief biographical notes on the contributors to the book then follow, with acknowledgements, a useful glossary of terms, endnotes, a reading list and index rounding things off.

How the book came together

The idea of a book was first suggested by Jeremy Toynbee of 5m back in March 2022. We were interested but so busy with filming we had no option but to make positive noises in reply and put it on the back burner. But he was persistent. Over the summer of 2023, when we had a little headspace, a plan was made and a small book team assembled.

Books are normally written by one, two, perhaps a few more, people, starting with an idea and working out from a plan. This book is different. It contains the voices of 22 people and we already had the content. Dauntingly it was spread over the 86 hours (5,171 minutes) of raw footage from which the final 90 minute film was crafted.

We handed the footage and transcripts, running to thousands of pages, over to the editing team of Molly, Priya and Jeremy. They weeded out all the 'ums', analysed the content and created a cohesive structure. Chapters 1, 2, 4–6 and 8 are formed from the interview transcripts worked into narrative form and imagined roundtable discussions with the experts. We wanted to maintain the energy of the spoken word and, especially in the three farmers' chapters their voices, and so they have only been lightly edited. Chapters 3, 7 and the interludes were commissioned specifically for the book.

A note on timescale. Filming took place from late 2021 through to early 2023 and the film focuses on 2022. We took the decision early on not to extend or update the story beyond then, for the main reason that Anna, Adrienne and Ben had already given so much time to the project that to ask them to do more was unconscionable.

A note on measurements. Metric units are used. Inches, feet, miles and acres are still much used in farming and have been kept with a metric conversation. Not least in the title! *152.4 Millimetres of Soil* is hardly catchy.

The film and book are closely related but stand on their own. With this book we have aimed to provide a slightly different perspective to the film. It doesn't need to be read from front to back; you can dive straight into any chapter – the story of one particular farmer, the details of one particular topic – but if you do choose to go from cover to cover then we hope to take you on a journey. One that starts with the soil, moving through the stories of farmers working to protect it and, finally, through to the voices of the experts we interviewed on where we go from here.

Soil is life.

It's a miracle, but that miracle has created a disaster

COLIN RAMSAY & JEREMY TOYNBEE

With contributions from
MIKE BERNERS-LEE, HENRY DIMBLEBY,
VICKI HIRD, SATISH KUMAR, TIM LANG,
NICOLE MASTERS, IAN WILKINSON &
DEE WOODS

The briefest history of soil and farming

oil. It's everywhere, it's right there under your feet. We barely give it any thought, but we really should; without soil there is no life.

Our ancestors did. It was part of their culture; they celebrated the land and its life-giving fertility. Crops were first cultivated about 12,000 years ago, followed by the domestication of animals, irrigation and the invention of the plough. Agriculture fundamentally changed the way we live, as we moved from nomadic hunter-gatherer existence to more permanent settlements.

It takes thousands of years to transform rocks, water and dying plants into biologically active soil. Its fertility, however, can be lost in mere decades. When soil became spent, civilisations would die, adapt or go to war to take that fertility from others. The early Britons and subsequent settlers all sought the fertility of British soils, expanding farming and the wool trade across these isles to feed and clothe their growing populations.

The British Agricultural Revolution significantly increased agricultural yield between the 17th and 19th centuries and was characterised by widespread changes in farming practices and the enclosures which forced many people off the land and into cities, which in turn helped drive the Industrial Revolution.

The British and other European empires exported this productionist mindset, focused on high-output, large-farm monocultures to the rest of the world. This caused huge damage to ecosystems, pushing indigenous people off their land and ignoring their age-old systems of sustainable food production well suited to their soils and climate.

Scientific and technical advances in the first half of the 20th century transformed farming through the 'Green Revolution'. Synthetic fertilisers, pesticides and government subsidies increased yields and farmers became heroes, feeding an ever-growing global population.

The Green Revolution – a miracle?

> The story of how the modern world feeds itself is a triumph of human ingenuity – but also of devastating unintended consequences.[1]

At the end of the Second World War there was real concern globally about how we were going to feed ourselves. The population stood at around 2.5 billion and was predicted to rise to over 8 billion. Food security made headlines and was extensively debated by government before and after the passing of the Agriculture Act 1947.

> [T]oday we are not able, with our home produce to feed more than half the population.[2]

THE SPECTRE OF FAMINE
From many sources evidence arrives that the war against famine is going against us.
Western Daily Press, Wednesday 15 May 1946

FOOD CRISIS
As Britain prepares for the biggest food drive in her history, warnings come of the return of famine conditions to Europe within a few months.
Western Times, Friday 29 August 1947

HOME FOOD OR FAMINE
[T]he Government will have to give practical proof without delay
that it means business. Hunger will be the penalty of failure.
Dundee Courier and Advertiser, Wednesday 3 September 1947 [3]

Historically, when more food was required, we simply turned more land
over to agricultural production as had been done during the war. Widespread ploughing up of scrub and grassland enabled Britain to increase
crop output and raise self-sufficiency from 30% to 80%.[4] However, in
the UK and globally stocks of productive land were running short. There
simply was not enough land to meet the demands of the predicted more
than threefold population increase:

> Whether we like it or not, there are only so many acres available
> for agriculture. Furthermore, that acreage is totally and utterly
> inadequate to feed the people of this country. … the fullest use has
> to be made of every acre.[5]

We needed new answers: they came in the form of chemistry, biology and
mechanisation.

Farmers had been aware of the benefits of enriching the soil with fertilisers
for thousands of years. Petrus de Crescentius' (c.1305) *Ruralia Commoda*
drew on Roman sources and was translated from its original Latin into Ital-
ian, German, French and later Dutch and English. Bernard Palissy writing
in 1563 stated, 'When you bring dung into a field it is to return to the soil
something that has been taken away.'[6] The chemist Humphry Davy, in
1813, published his bestselling, *Elements of Agricultural Chemistry*, having:

> analysed many organic materials, including blubber, bones, farmyard
> manure, fish refuse, guano, hair, horns, peat, night soil, seaweed,
> soot, straw and urine, to assess their value as soil amendments.
> He also analysed ammonium sulphate, common salt, gypsum,
> lime, magnesium limestone, peat ashes, potassium sulphate and
> sodium sulphate to determine their value. … His analysis of pigeon
> droppings led Davy to forecast that Peruvian guano would be of
> great benefit; guano was first imported in 1840 into Great Britain.[7]

South American supplies were of much higher quality than European sources, which were diluted by heavier rainfall. The late 19th century saw heavy extraction and the depletion of reserves built up over centuries. There was a scramble to take control of sources on islands across the Pacific Ocean and Caribbean Sea by the US, European and South American countries culminating in the Nitrate War (1879–83). Up to this point fertiliser predominately came from organic sources, although it was increasingly manufactured.

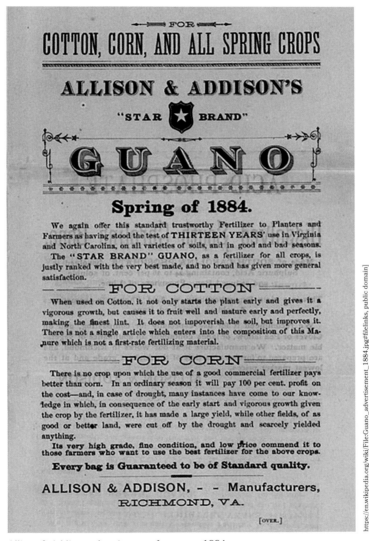

Allison & Addison advertisement for guano, 1884.

In response to increasing demand for nitrates and ammonia at the beginning of the 20th century German chemist Fritz Haber developed a synthetic process to convert atmospheric nitrogen (N_2) into ammonia (NH_3). Carl Bosch, working for chemical company BASF, was assigned to scale up Haber's laboratory success to an industrial scale, which he succeeded in doing in 1910. They were both later awarded Nobel Prizes (in 1918 and 1931, respectively).

The Haber–Bosch process became critical to Germany's First World War effort supplying the large amounts of nitrate needed for munitions production. The coming of peace released ammonia from the Haber–Bosch process to again be used in the production of synthetic fertilisers; use of which rose steadily through the interwar years.

Insecticides, developed from wartime research on nerve gases, gave us organochlorines and organophosphates (DDT being one of the most widely used). These synthetic fertilisers and insecticides, along with new herbicides, allowed us to 'control' nature and agricultural production rose.

After the Second World War, pioneering research by US botanist Norman Borlaug, who developed short-stemmed, high-yielding wheat and similar work elsewhere on rice and corn, brought further gains. These new strains responded well to and increasingly became dependent on synthetic chemical inputs. Developments in the size and efficiency of farm machinery, intercontinental transport and refrigeration powered by fossil fuels allowed food to be stored for longer and moved around the globe.

During the Second World War, the stranglehold of German U-boats on food imports to Britain had threatened to starve Britain out of the war and had exposed the fragility of our food system. Wartime government control of agriculture and food production was extended and developed into our modern subsidy systems (such as the Common Agricultural

SYNTHETIC FERTILISERS AND INSECTICIDES, ALONG WITH NEW HERBICIDES, ALLOWED US TO 'CONTROL' NATURE AND AGRICULTURAL PRODUCTION ROSE.

Policy in the EU). Subsidy systems paid farmers to produce ever greater quantities and fixed artificially high prices guaranteeing returns. Subsidies encouraged production, but ultimately led to damaging distortions in the market (see Interlude III, pages 110–113).[8]

How the Ferguson System fights hunger & poverty

The world's population is increasing faster than food production. And dear food—scarce food—is causing high living costs and world unrest. Slow, laborious farming with power animals or inefficient machinery can never solve this food problem. But a solution has been found—with the Ferguson System of *complete* farm mechanisation.

In 4 years Harry Ferguson Ltd., Coventry, have sold 150,000 tractors and the implements to work with them. These represent not merely new machinery but a new farming system. A system that is working successfully in 76 different countries.

The Ferguson System combines all the advantages of light and heavy machinery. It costs less to buy, less to run, and less to maintain. It

enables old men and women, boys and girls to do a strong man's work—faster, better, more cheaply than ever before. This system has already produced up to *ten times more food* in some areas. It is helping farmers *everywhere* produce more food at less cost from every available acre.

GROW MORE FOOD — MORE CHEAPLY — WITH **Ferguson**

Ferguson tractors are manufactured for Harry Ferguson Ltd., Coventry, by The Standard Motor Company Ltd.

XXIV

Harry Ferguson Ltd advertisement from Festival of Britain Guide, 1951: 'This system has already produced up to *ten times more food*'.

Chemistry, biology and mechanisation, coupled with state financial incentives, brought monocropping in ever larger fields, with bigger machinery and fewer people. We were in the age of industrial farming.

Population size, food volume and the area of land under cultivation had historically increased slowly together: more people, more food needed, more land turned over to agriculture. Industrial farming decoupled this historic relationship. From the middle of the 20th century the three lines diverge (Figure 1.1).

The Second Agricultural Revolution, beginning in the 1910s and becoming the Green Revolution of the mid-20th century as advances in the industrial North were exported to and forced on the South, did bring significant benefits.

- Increased agricultural productivity, particularly in the staple crops of wheat, rice and maize, helped address food shortages and hunger in many parts of the world.
- Higher agricultural productivity increased farm incomes and raised many people out of poverty at the individual level and created opportunities for countries to export surplus food, leading to increased income through international trade at the country level.

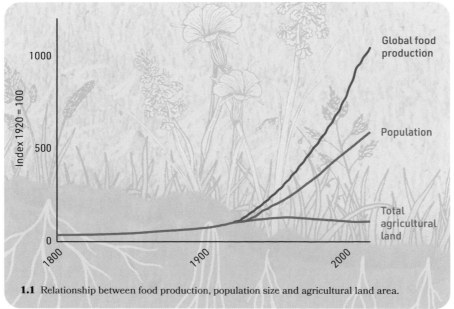

1.1 Relationship between food production, population size and agricultural land area.

The National Food Strategy, Fig. 3.1, p. 40.

- Innovation, transferred knowledge and technology between countries and regions allowed farmers to adopt more efficient and productive farming practices, including the development of new crop varieties, improved irrigation systems and the use of synthetic fertilisers and pesticides.

But this abundance of cheap food has come with huge hidden costs.

That miracle has created a disaster

There are almost eight billion humans alive on Earth – more than ever before and yet the threat of mass starvation has receded.[9]

IAN WILKINSON: It's astonishing how well we've done in that respect. But we didn't expect the consequences around biodiversity loss, nor did we expect climate change. I was certainly not taught about that when I was at agricultural college.

HENRY DIMBLEBY: It all came at a terrible cost. Farming is the single biggest cause of biodiversity collapse, of water pollution and scarcity, of deforestation, of the clearing out of our oceans of fish, and is the second biggest cause of climate change, and it is also now, by far, the biggest cause of avoidable disease.

TIM LANG: Literally, the land is burning. Literally, the icecaps are melting. Literally, soil is being washed into the sea. We're polluting rivers and seas and polluting our bodies, and nothing is done.

MIKE BERNERS-LEE: So now we're getting into a point where if we continue with the same practices, the whole thing's going to collapse. The soil systems are collapsing, the yields are going to go down, the biodiversity is haemorrhaging.

HD: So, it's got to change. The question is how much harm is done to our health and the environment before it changes? But literally it cannot survive.

TL: Society's got to have a very serious discussion. What do we want from our land and what are we prepared to pay for it? If we want to spend more money on a car sat outside our house than on our food. Well, so be it. Carry on doing what we're doing. But that will be disaster ecologically. That will be disaster in public health terms. It won't work in the long term. So, I think we're approaching a crunch point where we have to be very grown up about the solutions. It's partly about the land, but it's mostly about political economy and society. What do we want from our food system? Do we want food to be a key element in how we live and how we live well? Or do we want food just to be fuel and cheap?

HD: We looked at this disaster that followed the miracle and we tried to understand what was causing it in the *National Food Strategy*, which was an independent report for government, setting out how we could create a food system that fed us all nutritious food at a reasonable price without destroying our health and destroying the planet. The lens that we use to understand that was system dynamics, which looks at the feedback loops in complex systems and how they can go wrong.

All complex systems, whether it is a farm, a distribution network, the photosynthesis mechanisms of a plant, they all have certain characteristics and one is that if feedback loops go wrong, they can collapse without warning relatively quickly.

We identified two fundamental feedback loops that weren't working. One was the absence of a feedback loop. So, in the environmental space, nature is invisible in the way in which we measure human prosperity, human achievement. So, you can't count it in your wallet. It's not on the balance sheet for companies. It's not in the way we measure GDP.

And in fact, it's not only not there, governments worldwide subsidise the destruction of nature to the tune of US$500 billion a year in subsidies to big farming and to energy companies. So, we're actually not valuing nature. We're giving it a negative cost. We're paying people to destroy nature. And if we're going to solve the environmental side, we need to reverse that. We need to find ways of valuing nature and all of the systems that we use to prosper.

LITERALLY, THE LAND IS BURNING. LITERALLY, THE ICECAPS ARE MELTING. LITERALLY, SOIL IS BEING WASHED INTO THE SEA.

17

And the other mechanism that wasn't working was what we call the junk food cycle, which is the toxic interaction between our evolved appetite, which makes us seek out food that is highly calorie dense, that is high in sugar and fat and salt and low in fibre, and market forces. Food companies have worked out that it is easier to market that stuff to us than it is the healthy stuff. They have spent more money developing and marketing those foods. We eat more, they spent more, we've eaten more, and we've got sick.[10] That drive, that genetic drive is so strong that you are not going to be able to overcome that toxic feedback loop. It's called a reinforcing feedback loop technically. So those were the two insights. And then all of the recommendations in the report were centred around what policies are needed to introduce nature into the system, on the one hand, and to make it less attractive to sell and market things that kill us than it is to sell and market things that make us healthy, on the other.

TL: Capitalism has treated fossil fuels as never ending and has not internalised the costs of environmental damage. It's ignored the public health externalities of super abundant, fatty, salty, sugary, ultra-processed foods.

HD: To give an example of the way in which junk food dominates our diet, we spend in the UK £2.2 billion a year on fruit and vegetables. We spend on confectionery, which is one small category of junk food, £3.9 billion a year. Also, 70–80% of all advertising is on junk food. It is completely in one direction, and most of that junk food is made from highly refined wheat and sugar and vegetable fat. Very cheap commodity products are taken and turned into very palatable foods and end up in destroying both our health and planetary health.

MBL: So, I'm going to ask this question. Do we produce enough food? And if so, why are people still malnourished and hungry and still having problems accessing healthier food?

The first thing to say is that we already grow enough of all the human essential nutrients at the global level that we need to feed everyone alive. That's calories, protein, vitamins – the lot. We have enough of it. Our problem is sharing it around. The average human needs about 2,300 or so calories per day. And the great news is that we grow almost 6,000 calories per person per day of human digestible food. So that's nearly two and a half times what we need. Isn't that fantastic? How could anyone go hungry when that's the case? What's the core of the problem?[11]

NICOLE MASTERS: The amount of food waste is absolutely phenomenal. We've got to take care of food waste, not drive more and more production of less and less quality. If we look at the nutrient density of our food, we know that nutrient levels may have dropped by as much as 46%.[12]

MBL: Let's follow the food that we grow on its journey from the field to our mouths and other places. So, first of all, we lose a little bit in harvest losses and in storage and so on. And that's a problem, but it's

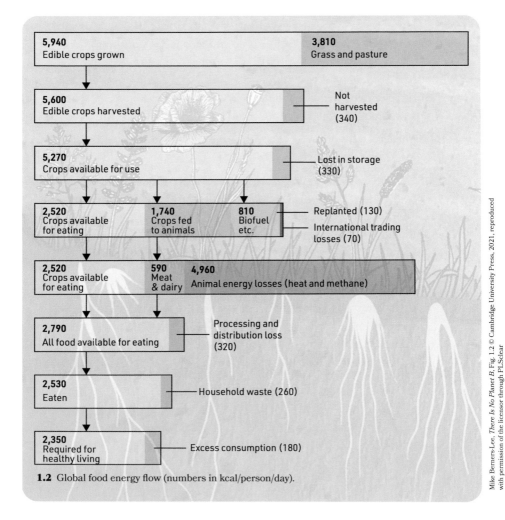

1.2 Global food energy flow (numbers in kcal/person/day).

Mike Berners-Lee, *There Is No Planet B*, Fig. 1.2 © Cambridge University Press, 2021, reproduced with permission of the licensor through PLSclear

not a showstopper. But the real problem is that when we get all this food what we do with it. And the biggest waste in the system, if you like, or loss in the system, is that we feed so much of it to farm animals.

TL: Using land to grow grains to feed cattle, it's crazy.

MBL: Yeah, it's not something we see happening when we look at the fields. They're either full of human edible crops or they look as though they are, or else they're full of cows eating grass. But we don't see the fact that there's soya plantations all over the world on deforested land that's had to be cleared in order to make way for plantations of crops to feed cows.[13]

The problem with feeding human digestible food to animals is it introduces a huge inefficiency into the food chain. It's far more efficient to get human digestible food and feed it straight to humans. When you give it to an animal, the animal will typically give you back something like 1/10th of the nutrition, because it wastes calories by doing inefficient things like walking around, keeping warm and sometimes burping up methane.

TL: There are very big issues around land use. Why aren't we growing more horticulture and fewer animals?

MBL: So historically, meat was an aspirational thing for health reasons. And then we went through a phase where it started becoming affordable for more and more people and everybody started getting used to having it every day. By eating meat, you got to live longer and healthier lives. Now, that's not the case anymore. In fact, the opposite is more likely.

So the science is clear whether you care about climate change or biodiversity or feeding the world, we need to reduce by a long way the amount of meat and dairy in the global diet, which means that in the UK consumption particularly needs to come down because we already eat so much of it. We don't have to be extreme about it.

WE GROW ALMOST 6,000 CALORIES PER PERSON PER DAY OF HUMAN DIGESTIBLE FOOD. HOW COULD ANYONE GO HUNGRY WHEN THAT'S THE CASE?

I don't think any of us need to go vegetarian or vegan unless we want to. But we need a lot of dietary change.

If we make the right dietary changes now, specifically towards less meat and dairy, there are opportunities for us to live longer and healthier lives, to feel better about our lives, partly because the food will work better in our bodies, and partly because we'll feel better about what we're eating because we know it's better for the planet.

There are some other big problems as well, such as non-food uses, mainly biofuel. If you look at the amount of energy we get in return for the amount of food we have to sacrifice, it's an incredibly steep trade off. So enough food to give me all the calories I need for a day would only create enough biofuel to drive a very efficient small car for about 1.5 km: 1.5 km against a day's worth of calories.

DEE WOODS: The system isn't a broken system; it's working exactly how it's meant to be. It's a system designed to extract and build profit. Rather than to ensure that everyone has equal access.

TL: Supermarkets are very efficient. They deliver what people want. People can go to a hypermarket and they have 35,000 different items to choose from. This is consumer capitalism at its absolute best. It is. They are utterly amazing, brilliantly run, really efficient. But until you deal with the corporate power of nine retailers who have 94.5% of food retail sales in Britain, you're not going to have a level playing field. The retailer is king, the retailer is sovereign. Farming is squeezed. Britain spends about £225 billion a year on food, of which farming gets about 8%.

We've got to think about a different model of getting food to people. The crisis is how we run an economy. That is where the problems start. How much money are farmers, primary producers,

IT'S A SYSTEM DESIGNED TO EXTRACT AND BUILD PROFIT. RATHER THAN TO ENSURE THAT EVERYONE HAS EQUAL ACCESS.

fisherfolk, etc., and horticulturalists going to get? Are consumers going to pay the full cost?

SATISH KUMAR: We think that food should always be cheap. We don't pay the just price for the food, so we are prepared to pay lots of money for cars, computers, houses, airplanes, everything else. But food must be cheap.

SK: All the animals, bugs, insects and even trees are fed by the six inches of soil. It's a miracle.

VICKI HIRD: What I find absolutely extraordinary and what we're really only just uncovering is the amazing way the soil works. It's a whole community below ground with worms, springtails, beetles, spiders, working with the fungi and the rhizosphere with the plants, with the roots, and with the physical chemistry. It's just extraordinary down there.

NM: Very much so, but there are serious issues with soil structure, with soils not breathing and with compaction. Soils are not breathing. A soil that doesn't breathe is totally in dysfunction and dysbiosis. If we think about soil as like our gut systems. These soils have Crohn's disease, rampant diarrhoea or constipation.

SK: We humans don't understand soil. We think that working on the land is something backward. To be educated, be smart, be clever you don't work on the land. You work in industry, or you become a lawyer, an accountant, a civil servant.

People are referred to as consumers. This is totally wrong. We are not consumers. We are makers. Each and every human being is capable of making. Let's bring back the culture of making, not consuming. Agriculture should be for everybody. All those who eat must participate in growing food.

NM: I think in many ways society's lost its ability to think through complex questions. The biggest hurdle I see is what I call the top paddock – our ability to think cognitively and complexly. Right now, farmers are under so much stress and there's so much debt, there's so much external pressures, there's legislation, there's climate change, there's a separation of urban and rural communities and people are literally so stressed. We find that cognitively the ability to deal with anything new really shuts down because we're literally in a fight or flight response. Farmers are just doing what they can to survive. And so, it makes it really challenging to take on new information and new ideas. People are thinking, 'I'm putting one foot in front of the other and I'm going to do what I've always done because literally that's all I can deal with.'

IW: When farmers and growers come to us at FarmED, they often ask the same questions: Will this new farming system work? Is it reliable? Will it make money? And what do I have to do to change to it? And so, there are a number of barriers. There's the economic barrier, and there's the knowledge barrier. The biggest barrier, perhaps, is the market. They're asking: If I grow something new, a whole range of diverse crops, for example, who's going to buy them? Is there a market for it?

AGRICULTURE SHOULD BE FOR EVERYBODY.
ALL THOSE WHO EAT MUST PARTICIPATE IN
GROWING FOOD.

23

The briefest summary and a glimmer of hope

Machinery has replaced people on the land, and hedgerows, meadows and trees have been torn down to plant vast, chemically dependent monocrops. Hunger for cheap meat and dairy has intensified livestock production, pouring effluence and chemicals into our streams and rivers. Enormous tractors plough deep into the ground, degrading and compacting the soil. Bare and unprotected the soil is eroded, washing away into our rivers and seas. The result … Britain takes the crown as one of the most nature-depleted countries on Earth.

The irony is we have endless food choice in our corporate supermarkets, but little influence over where food comes from, how it's produced or where to find nutritious, affordable food that's grown in a truly nature-friendly way.

Yet change is in the soil. Regenerative agriculture offers a suite of farming practices that rebuild soil organic matter and restore biodiversity. This is part of a growing movement of agroecological farmers and food producers who are rejecting the industrial model and choosing to work with nature, not against it.

Now is the time to reimagine our relationship with the soil. To recognise how it feeds us, protects us and sustains us. We need to hold the living soil in our hands and feel the earth beneath our feet once more.

NOW IS THE TIME TO REIMAGINE OUR RELATIONSHIP WITH THE SOIL. TO RECOGNISE HOW IT FEEDS US, PROTECTS US AND SUSTAINS US.

Regenerative farming, what is it?

MARINA O'CONNELL

There are two different agricultural timelines: conventional/ industrial and sustainable/ecological. They diverged in the early 1900s, when nitrate fertilisers, pesticides and herbicides began to be used widely. At that point, much of farming went down the industrialised route, which was slowly rolled out from Europe and around the globe (Figure 2.1).

There have been a series of ecologically or biologically based farming systems that have responded to the roll out of industrial farming. The first response was biodynamic farming, and then came organic farming in the late 1930s. Permaculture, agroforestry and agroecology arrived during the 1970s, and then finally, we have regenerative agriculture. So, what do these terms mean?

Biodynamic

At its simplest, I describe biodynamic as organic with bells. It was the first form of biological or ecological farming and was developed in 1924 by Rudolf Steiner, a scientist and philosopher. It came about when farmers attending a conference in Koberwitz (a small village in then Germany, now Poland) asked him for guidance on how to farm without nitrate fertilisers. They had been using industrial chemicals and fertilisers for about 10 years, and they'd noticed a drop off in the fertility on their farms.[1]

Marina O'Connell, *Designing Regenerative Food Systems* (Hawthorn Press, 2022), reproduced with permission.

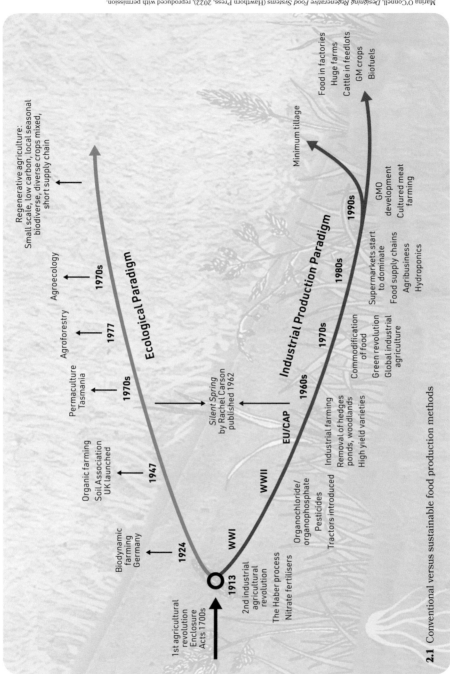

2.1 Conventional versus sustainable food production methods

Steiner came up with a set of practices and principles that became biodynamic farming. It is almost identical to organic farming, which developed from it. There are no pesticides, no nitrate fertilisers and very high welfare standards. Most biodynamic farmers try to sell their produce locally and directly to their customers within a community supported agriculture system.

A biodynamic farm should:

1. become an organism with closed-loop systems
2. work with nature rather than against it
3. create biological cycles and avoid pollution
4. grow and produce high-quality food that is sold as locally as possible
5. connect plants, animals and people to cosmic and planetary movements, and work with them
6. promote biodiversity from the soil up
7. be socially and community connected
8. be economically viable
9. allow animals to express their nature, to ensure their high welfare
10. enhance the landscape.[2]

Biodynamic farming is accredited by the International and European Regulations for Organic and Biodynamic Methods and regulated by regional associations around the world. Biodynamic food is sold under the trademark of the Biodynamic Federation – Demeter.

One of the main differences between organic and biodynamic systems is that biodynamic uses preparations of fermented manures and plants, which are applied to the soil and crops. A little bit like the current use of compost teas.

THEY HAD BEEN USING INDUSTRIAL CHEMICALS AND FERTILISERS FOR ABOUT 10 YEARS, AND THEY'D NOTICED A DROP OFF IN THE FERTILITY ON THEIR FARMS.

When organic farming, in part, arose out of biodynamic farming, early practitioners removed these preparations. Why? Organic farming developed through the late 1920s and early 1930s when agriculture was transitioning into industrial farming, and using herbal plants and fermented manures seemed a little bit 'witchy'. Today, we often use the word witchy or hippie as a derogatory term, but it's basically our indigenous European methods of farming, which we are now learning to re-embrace.

Organic

People are more familiar with organic farming: no pesticides, no herbicides and nitrate fertilisers replaced with biological sources of fertility. The organic movement arose in the 1920s from observations of indigenous farming practices in China, Japan, Korea and India,[3] which were adapted for use by key American and British adopters. The Soil Association was founded in the UK in 1946.

Organic farming strives to meet four key principles.[4]

- The Principle of Health – organic agriculture should sustain and enhance the health of soil, plant, animal and human as one and indivisible.
- The Principle of Ecology – organic agriculture should be based on living ecological systems and cycles; it should work with them, emulate them and help sustain them.
- The Principle of Fairness – organic agriculture should build on relationships that ensure fairness with regard to the common environment and life opportunities.
- The Principle of Care – organic agriculture should be managed in a precautionary and responsible manner to protect the health and well-being of current and future generations and the environment.

Many organic farmers try to be closed loop, but they don't actually have to be. You could still, for instance, go into a supermarket and buy some organic blueberries flown across the world from Argentina, wrapped in plastic. A lot of organic farmers don't produce like that, but it is possible in an organic system. So organic is still slightly more linear, it can be a bit more like the industrial mindset; but with the artificial inputs removed.

Organic farms have to meet the standards and regulations of the International Federation of Organic Agriculture Movements (IFOAM

– Organics International). The regulations are administered by regional associations, such as the Soil Association (see page 279) or Organic Farmers & Growers (OF&G, see page 278) in the UK. Organic produce can then be sold under trademarked logos.

Permaculture

Permaculture, fusing the words 'permanent' and agriculture', was developed in Tasmania by Bill Mollison and David Holmgren. They came up with a very powerful design methodology to create ecologically harmonious landscapes for people and other non-human beings and nature. Permaculture design centres on zones ranging out from the farmer (0) to the most distant and least visited (5), into which the desired elements on the farm are arranged depending on the energy and frequency of visits needed to maintain them.

Permaculture systems aim to be closed loop and generally contain more perennial crops than other farming systems. Mollison developed a set of key principles (emboldened in the following summary).[5] Nature he argued is powerful, and we should **work with nature rather than against it**, taking a creative approach to problem-solving that fits a solution to the circumstances on the farm so that **the problem is the solution**. We come to those solutions by observing, trialling and adjusting, to ensure we **make the least change for the greatest effect**. Crops and animals should be **stacked** so that **everything gardens** with species in **cooperation and not competition**. The resulting **yields are limited only by our imaginations**, especially when the definition of yield is broadened to include biodiversity gains, mental and physical well-being and carbon sequestration, and when **pollution is considered to be a wasted resource** to be designed back into a circular closed system. **Biological resources**, that are local and natural, should be preferred and **diversity**

YIELDS ARE LIMITED ONLY BY OUR IMAGINATIONS, ESPECIALLY WHEN THE DEFINITION OF YIELD IS BROADENED TO INCLUDE BIODIVERSITY GAINS, MENTAL AND PHYSICAL WELL-BEING AND CARBON SEQUESTRATION.

should be maximised, especially at the **edges of the system** where two ecosystems interact. Planning should **create patterns, then design from patterns to details**, making **efficient use of energy across zones, sectors and slopes**. Planning should also allow for **evolution of the system** and **build in resilience,** with every element performing multiple functions. Permaculture is a central concept of the wider regenerative agricultural movement.

Agroforesty

Agroforestry, which as its name suggests is a combination of agriculture and forestry, has ancient roots and is practised worldwide in much the same forms. The term was coined in 1977. Here we are talking specifically about its rediscovery and reintroduction into the Global North's agriculture systems. These are traditional systems that were literally and figuratively ripped out by industrial farming. Now we are putting the trees back into the system.

Modern agroforestry often features in permaculture design and in biodynamic, organic and regenerative farming systems. Trees help with carbon sequestration and water management and with pest and disease control. Trees provide animal fodder and shade; they build diversity and work in symbiotic relationships with the crops planted between and among them. Of course, quite a lot of the trees we use, we can crop as well – stacking crops and enterprises on the same land increases the combined yield.

Agroforestry has the two main principles of increasing both biodiversity and multifunctionality, which together improve resilience in the food system. In practice it offers a range of implementation possibilities (Table 2.1). (See Interlude V, pages 212–217.)

TRADITIONAL SYSTEMS THAT WERE LITERALLY AND
FIGURATIVELY RIPPED OUT BY INDUSTRIAL FARMING.
NOW WE ARE PUTTING THE TREES BACK.

LOCATION OF TREES	AGROFORESTRY SYSTEM	PRACTICE BY LAND-USE CLASSIFICATION	
		FOREST LAND	AGRICULTURAL LAND
Trees within fields	Silvopastoral	Forest grazing	Wood pasture Orchard grazing Individual trees
	Silvoarable	Forest farming	Alley cropping Alley coppice Orchard intercropping Individual trees
	Agrosilvopastoral	Mixtures of the above	
Trees between fields	Hedgerows, shelterbelts and riparian buffer strips	Forest strips	Shelterbelt networks Wooded hedges Riparian tree strips

Table 2.1 Types of UK agroforestry.[6]

Agroecology

Agroecology was a term coined in the 1940s, which came to the fore in the 1970s and 1980s: agroecology, a fusion of agriculture and ecology to create a system of ecological farming. Unlike biodynamic and organic, there is no codification, so there's wider take up of agroecological farming systems worldwide.

As a movement it's completely opposed to the industrialisation of food, to transnational companies and agribusiness, and to patriarchy. La Via Campesina,[7] the body that since 1993 has overseen agroecology, defines the movement as anti-capitalist, and it actively campaigns to achieve a wholly different system. And in fact, I think it's arguably one of the largest social and political movements in the world.[8]

> We call on our fellow peoples to join us in the collective task of collectively constructing agroecology as part of our popular struggles to build a better world, a world based on mutual respect, social justice, equity, solidarity and harmony with our Mother Earth.[9]

Agroecological farming has pretty much the same principles and practices that we find in organic systems, but again much more emphasis on the closed-loop concept: the idea of not buying anything in and produce being sold locally. As a point of differentiation, the concepts of food sovereignty, equitable access to land and natural resources, peasants' rights, dignity for migrant and waged workers, and international solidarity are absolutely core to agroecology. The idea is that everyone has a human right to grow and eat their own food or food that is produced locally.

Being a wide-ranging, encompassing movement, definitions of agroecology and its principles vary from the 9 key strategies of La Via Campesina to FAO's 10 Elements of Agroecology[10] and the HLPE's expanded list of 13 agroecological principles (Figure 2.2).[11] The latter of which is listed here.

1. Recycling.
2. Input reduction.
3. Soil health.
4. Animal health.
5. Biodiversity.
6. Synergy.
7. Economic diversification.
8. Co-creation of knowledge.
9. Social values and diets.
10. Fairness.
11. Connectivity.
12. Land and natural resource governance.
13. Participation.

EVERYONE HAS A HUMAN RIGHT TO GROW AND EAT THEIR OWN FOOD OR FOOD THAT IS PRODUCED LOCALLY.

FOOD SYSTEM

11 CONNECTIVITY
Ensure proximity and confidence between producers and consumers through promotion of fair and short distribution networks and by re-embedding food systems into local economies.

12 LAND AND NATURAL RESOURCE GOVERNANCE
Recognise and support the needs and interests of family farmers, smallholders and peasant food producers as sustainable managers and guardians of natural and genetic resources.

13 PARTICIPATION
Encourage social organization and greater participation in decision making by food producers and consumers to support decentralised governance and local adaptive management of agricultural and food systems.

10 FAIRNESS
Support dignified and robust livelihoods for all actors engaged in food systems, especially small-scale food producers, based on fair trade, fair employment and fair treatment of intellectual property rights.

9 SOCIAL VALUES AND DIETS
Build food systems based on the culture, identity, tradition, social and gender equity of local communities that provide healthy, diversified, seasonally and culturally appropriate diets.

AGROECOSYSTEM

8 CO-CREATION OF KNOWLEDGE
Enhance co-creation and horizontal sharing of knowledge including local and scientific innovation, especially through farmer-to-farmer exchange.

7 ECONOMIC DIVERSIFICATION
Diversify on-farm incomes by ensuring small-scale farmers have greater financial independence and value addition opportunities while enabling them to respond to demand from consumers.

6 SYNERGY
Enhance positive ecological interaction, synergy, integration, and complementarity among the elements of agroecosystems (plants, animals, trees, soil, water).

5 BIODIVERSITY
Maintain and enhance diversity of species, functional diversity and genetic resources and maintain biodiversity in the agroecosystem over time and space at field, farm and landscape scales.

4 ANIMAL HEALTH
Ensure animal health and welfare.

3 SOIL HEALTH
Secure and enhance soil health and functioning for improved plant growth, particularly by managing organic matter and by enhancing soil biological activity.

2 INPUT REDUCTION
Reduce or eliminate dependency on purchased inputs.

1 RECYCLING
Preferentially use local renewable resources and as far as possible, close resource cycles of nutrients and biomass.

2.2 The FAO High Level Panel of Experts's 13 agroecological principles.

Regenerative farming

Finally, we come to regenerative farming. We always used to talk about sustainable farming, but sustainable is no longer good enough because our soil has become so degraded. It has been reported that globally we've only got 60 harvests left in our soils.[12] So, what we actually have to do is regenerate our soils, we have to take them from a poor place to a better place. Sustainable suggests carrying on – but now we have to regenerate, we have to repair the damage.

Regenerative farming is a term that I take to mean regenerating soil in particular. When you regenerate the soil, you kick off **five cycles**. You regenerate the soil biome, the biological aspects of the soil, which starts **nutrient cycling**: we generate nutrients, we don't need to bring fertilisers in. This means the amount of organic matter in the soil goes up and thereby holds on to the water, so we regenerate the **water cycle**. Healthier soil, rich in organic matter, sequesters more carbon, so we regenerate the **carbon cycle**. Soil is the bottom of the biodiversity food chain, so once we have healthier soil we get more bacteria and fungi, we then get more worms, more beetles, more birds and so on up the food chain – this is the **biodiversity cycle**. And finally, hopefully we start to regenerate our economic and social ecosystems as well. As the soil recovers, the farms and farmers themselves often change. Sales tend to become more localised, so the farmer derives more of the income and connects better with their customers. Farms often become mixed, or more mixed, and require more labour, so the farm becomes populated with many more people – workers and visiting customers alike. Charles Massy calls this the **economic and social cycle**.

So we're using multiple farming methods to create the most regenerative farm that we possibly can. There are many tools that we can use to repair our soil and then kick off those five cycles.

There are five generally accepted regenerative farming principles (Figure 2.3), introduced to many (including our own Anna and Andrew Jackson, see Chapter 4) by Gabe Brown, among others, through his extremely

SUSTAINABLE SUGGESTS CARRYING ON — BUT NOW WE HAVE TO REGENERATE, WE HAVE TO REPAIR THE DAMAGE.

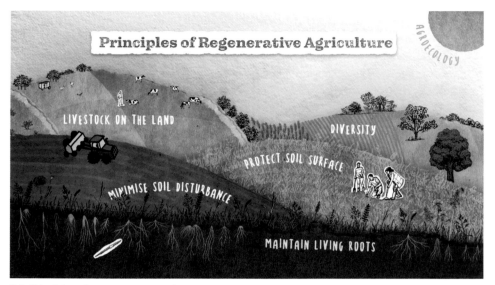

2.3 Principles of regenerative agriculture.

influential book, *Dirt to Soil*. They are not prescriptive nor designed for any particular system. They do not all have to be done together and not all will suit each locality. Proponents of regenerative agriculture often talk about their journey, starting with one change and building from there as their understanding and confidence increases.

1. Minimise soil disturbance

Soil supports a complex network of worm-holes and fungal hyphae and a labyrinth of microscopic air pockets surrounded by aggregates of soil particles. Disturbing this by regular ploughing or heavy doses of fertiliser or sprays will degrade the soil.

2. Protect (armour) the soil surface

Keeping the soil surface covered with living plants or crop residues protects the soil, and the organisms living within it, against weathering and erosion damage. The most successful way of implementing this principle is through over-winter cover cropping, alongside judicious maincrop and cover crop management. A good soil cover also prevents overheating during periods of hot sunshine, or freezing in winter, both of which are antagonistic to healthy soil.

35

3. Maintain living roots in the soil

Having living roots present in the soil for as much of the year as possible keeps the underground ecosystem functioning. Plants remove CO_2 from the air and turn it, via photosynthesis, into sugars, which are the building blocks they need to grow. Up to 70% of these sugars are exuded directly into the soil to feed bacteria and fungi that cluster around the plant root, exchanging other nutrients and water for the exuded carbohydrates.

Use of cover crops is an effective way to ensure soils are covered and living roots are present at times when the soil may otherwise be left bare in a conventional arable system.

4. Maximise plant diversity

Monocultures do not happen in nature, and our soil creatures thrive on variety. Companion cropping (two crops grown at once and separated after harvest) and break crops (a crop grown in a rotation as a break from the main cereal crop) can be successful. Cover cropping (growing a crop which is not taken to harvest but helps protect and feed the soil) will also have the happy effect of capturing sunlight and feeding that energy to the subterranean world, at a time when traditionally the land would have been bare.

More conventionally, robust crop rotations ensure healthier soil and reduced weed and disease pressure. Rotations have a long history, culminating in the 18th-century Norfolk four-course system (typically, wheat, turnips, barley and a clover/grass ley).[13] There is also potential for growing crops through a living mulch of clovers which stay close to the ground and allow the cereal to tower above and be harvested when ripe, leaving the understory to carry on feeding the soil and fixing nitrogen.

5. Livestock on the land

This principle recognises the importance of responsibly grazing livestock to spread organic matter and increase nutrient cycling and plant growth. This can be done through high impact mob grazing: short disturbance followed by long recovery periods. Although for some farming systems the integration of livestock is not possible, this principle can still be considered, for example, through the use of manure and slurry as an organic source of nutrients.

The presence of livestock in a farming system, if done wisely, supports the regeneration of soil. Potentially the trillions of living creatures present in soil can vastly multiply their numbers when grazing livestock are incorporated into the system. As well as improving soil biodiversity, farm income can also be diversified with the addition of livestock enterprises. A diversity of farm animals (such as cows, sheep, chickens, pigs and goats) will further boost soil fertility and animal health.[14]

More recently a sixth principle is commonly added of **knowing/working within your context**, which was already there implicitly in many descriptions, including Gabe Brown's, but is now explicitly drawn out. Implementing the other five principles without observing and understanding the individual characteristics of a farm system, even just a field or a portion of a field, will impede success. Every farm needs its own transition plan, needs to monitor results and needs to adjust implementation in response.

I often think of farming as like a continuum. At one end we've got really industrialised farmers and at the other end we've got regenerative farmers. The hope is that everybody is moving along this spectrum to become more regenerative. Farms that are industrial might make the first step on the regeneration path by moving to minimum tillage. They might still use nitrate fertilisers or glyphosate, but they will use less of it, and the soils will start to sequester more carbon. While an organic farmer may think, 'Well, I'm going to start using mob grazing' or 'I'm going to put agroforestry rows on my organic farm.'

Hopefully we're all on a journey to become more regenerative. Regenerative agriculture is quite an inclusive term. I feel it doesn't exclude anybody, and it's quite easy to start, to begin the journey with one tool and then add others.

There are certainly barriers to transitioning, such as training and convention. I did a degree in horticulture and was taught in the industrial methods, which were called conventional, although I don't think there's anything conventional about conventional farming as it's only about

EVERY FARM NEEDS ITS OWN TRANSITION PLAN, NEEDS TO MONITOR RESULTS AND NEEDS TO ADJUST IMPLEMENTATION IN RESPONSE.

three generations old. Regenerative farming is not going backwards to pre-industrial farming. Not at all. We're using a lot of modern science and modern understanding of ecology and soil ecology in particular.

A major problem is the average age of farmers (around 50 or 60), which means they all would have been trained in the industrial methods. Making the switch to regenerative methods is quite a transition, and most farmers have no knowledge of how to do that, nor is training readily available. It can be quite nerve wracking because, when you start, your farm can go into a bit of a wobble until the ecology kicks back in again; your systems can go haywire, a bit like weaning yourself off a high-sugar diet. So quite often you take the first few steps, and then you get a huge pest or disease outbreak or a weed problem. They will rectify themselves, but you have to really hold your nerve.

This chapter was developed from an interview with Marina O'Connell with short additions from her book, Designing Regenerative Food Systems, *which is strongly recommended as a source for more detailed explanation of these systems and how to use them.*

Interlude I

Are we using our farmed land wisely?

VICKI HIRD

It's all about land. What we use, or don't use, it for. Who owns the land is also a significant part of a highly politically charged discussion. There is no question we need to change how we manage land, especially rural and farmed land, here in the UK and globally.

Land use matters for four key reasons: we need to eat, and eat healthily; we need to protect and restore nature and ecosystems; we need to tackle climate change through capturing carbon in nature; and finally, we need to manage all the other competing land uses from housing to recreation. We are not doing any of these well or in ways that support healthy rural livelihoods and communities.

An urgent case for action

The 2023 State of Nature report highlighted that much of our wildlife in the UK remains in serious trouble.[1] Nature provides so many services, including healthy soil created by the fungal, worm and other invertebrate systems; crop pollination by bees, moths and many other invertebrates; and pest and weed control by beetles and birds.

More erratic weather, flooding, droughts and changes in pest and disease threats are already happening. Climate change is already with us, and we are, sadly, unlikely to keep to below 1.5°C warming, so land use needs to adapt.

There are many other competing demands on land, including industrial developments, infrastructure, greenspace and recreation, energy production such as biofuels or solar farms, and housing and water management. Growing divisions over how we do all this are getting in the way of developing effective policies and market responses. One of the key divisions is whether we divide the farmed land up differently, to make more of it entirely available for nature and intensify production on the rest, or we manage both outcomes on the same land: the land sparing versus land sharing debate. There are also innovations and technofixes which may or may not play a role.

What are land sparing and land sharing?

Land sparing 'involves restoring or creating non-farmland habitat in agricultural landscapes at the expense of field-level agricultural production. Note this approach does not neces-sarily imply high-yield farming of the non-restored, remaining agricultural land'. The Intergovernmental Sci-ence-Policy Platform on Biodiversity and Ecosystem Services (IPBES), who provided this definition, is careful in providing the two-sentence clarifi-cation – it means taking land out of production to restore nature but does not automatically mean intensifying production on what is left. Others are less specific and define it as 'produc-ing more from less' to free up land for nature, which rings alarm bells.

Advocates of land sparing include those who are highly concerned by the loss of nature from key areas, such as forest, peatlands, wetlands and mead-ows. Their motivation is to restore that nature. Some detailed modelling suggests land sparing can work, but in common with all responses, we need to change what we demand of land and what we eat.[2] The evidence is overwhelming – that we do need to restore land just for nature – but the questions of how much, where and

what happens to the non-restored land still need answering.

A land sparing scenario, as IPBES and others suggest, must take a whole-system approach.[3] The land outside the spared 'reserves' must also be well managed for nature, or we risk weakening many benefits of the spared land. Farmed land and urban and peri-urban spaces must be nature friendly too, to provide the vital corridors and viable green spaces for nature to shelter in, move through and reproduce in. Some land sparing advo-cates suggest a three-compartmental model with land entirely for nature, lower-yielding farmed land and land for high-intensity food production.[4]

Land sparing advocates include some who have a vested interest in maintaining chemical, fossil-fuel powered and highly specialised, monoculture-based farming. They obviously want to protect their profits, so they argue that using these inputs 'carefully' to spare land, combined with new technologies, will free it up for nature – by growing more on less, or 'eco-intensification'. They are strong advocates against organic and other systems that yield less and differently from the land.

Alternatively, a land shar-ing approach is one in which wildlife-friendly farming systems are supported to deliver multiple benefits for nature and society while producing food. This is the kind of multifunctional land use that agroecological farming delivers, incorporating principles of diversity and low inputs with issues beyond the farmgate, such as fairness and quality research (see Chapter 2, page 31). Critically, land sharing can ensure more connectivity between land that is managed in an environmentally friendly way and protected 'spared' areas, so that these vital ecosystems interact and species can migrate.

The UN Food and Agriculture Organization, in setting out its 10 elements of agroecology, says:

By enhancing their autonomy and adaptive capacity, agroecology empowers producers and communities as key agents of change. Rather than tweaking the practices of unsustainable agricultural systems, agroecology seeks to transform food and agricultural systems, addressing the root causes of problems in an integrated way and providing holistic and long-term solutions.[5]

Agroecological farmers work to deliver better nature outcomes, providing a whole-system approach. Such land sharing makes huge sense as it produces enough food, more diverse food and works with nature to deliver nutrients and other services. If used everywhere, like all scenarios, it would require a sea change in what we eat and demand from land, and the amount we waste.

Is regenerative farming a form of land sharing? It can be, and for many regenerative farmers it is. In the last 10 years, regenerative agriculture (regen ag) has gained a higher profile and a far bigger following – as an approach to farming that, in theory, allows the land, soil, water, nutrients and natural assets to regenerate themselves, as opposed to conventional approaches to farming that can deplete these natural resources. There are, as yet, no set standards and a huge amount of hype, so it's hard to say whether regenerative farming is really sharing the whole land.[6]

Can innovation and technological fixes help?

Many look to technology for solutions. Innovations are designed which reduce pressure on land by decoupling food production altogether from land and soil or reducing how much land is needed.

Precision fermentation is of the most controversial. It involves producing some or much of our protein requirements in factories, using bacteria that feed on hydrogen and methanol – so no land is involved at all. George Monbiot makes much of precision fermentation as a solution in his book *Regenesis* and argues for a greatly reduced agriculture sector. One estimate suggests that precision fermentation using methanol needs 1700 times less land than the most efficient agriculturally produced protein, soy.[7] Others, like Jyoti Fernandes of the Landworkers' Alliance, dispute such a route for humanity for various reasons, including the need for productive mixed farming to maintain soil fertility and the need to support agroecological solutions which deliver for farmers, nature and people.[8]

Another hot technology is vertical farming which grows plants inside and with an extremely low land-take using soil-less hydroponic systems (water provides all the required nutrients) and artificial light.

Both precision fermentation and vertical farming have attracted huge investment over the past two decades. Some companies have since folded, and the realities are that the huge promises have yet to be realised, although they probably have some positive part to play in a future scenario especially in soil-depleted areas.

Genetically modified (GM) plants have always been strongly promoted by those seeking to gain from the technology. There is much hype about how these will reduce chemical use, provide disease resistance and deliver huge savings for farmers. While the main commercially available GM plants are herbicide tolerant, allowing spraying indiscriminately, we have yet to see the major gains in productivity, drought tolerance and disease resistance that have been promised.

Small-scale robotics, precision farming and artificial intelligence (AI), to enhance animal health and welfare, could have much to offer. For instance, renewably powered, low-impact robotic machinery could carry out weed control and plant nutrition. If they can be implemented without saddling farmers with further debt, they may provide valuable additional tools.

While we will inevitably see some of these technologies take hold, and beneficially so in some cases, the reality is they don't fix a broken food system and don't tackle unsustainable demand. They are too narrow in their use, and we know we have multifunctional needs from land and farming.[9]

The future is not compartmentalised

So which is best: sparing, sharing or innovating? The truth is that what we need is likely to be far messier than modellers and research scientists identify and far more complex than policy lobbyists like to portray.

From the evidence so far, and taking a multiple-outcomes perspective, I'm inclined to think we should largely aim for land sharing on farmed land, with some areas newly spared for nature: maybe the 30% of land spared, as agreed by nations at COP15. Innovative approaches should and may replace some of the worst excesses of the current system. Crucially, in all scenarios, we have to tackle food loss and waste – including the waste of crops diverted to biofuels, to feed industrially farmed livestock and to produce unhealthy, ultra-processed foods.

The science of soil

Soil health, carbon sequestration and the environment

PRIYA KALIA, HANNAH JONES
& LIZZIE SAGOO

Consider what each soil will bear, and what each refuses.

VIRGIL
Ancient Roman poet

Understanding soil

You may not have previously given it much thought, but soil is one of the most critical components of our living planet and the foundation of all Earth's ecosystems. It combines minerals with organic matter as well as water, air and lots of different types of micro-organisms, and can take thousands of years to form naturally.

We are only just starting to appreciate the amazing complexity of the world beneath our feet, and there is still so much more yet to learn. Soil is alive. A teaspoon of healthy soil will contain several billion different organisms, ranging from microscopic bacteria to earthworms or beetles. Author and researcher Merlin Sheldrake wrote in his groundbreaking book, *Entangled Life*:

> The view of plants as autonomous individuals with neat borders is causing destruction. If we follow the tangled sprawl of mycelium

that emanates from its [a plant's] roots, then where do we stop? Do we think about the bacteria that surf through the soil along this slimy film that coats roots and fungal hyphae? Do we think about the neighbouring fungal networks that fuse with those of our plant? And perhaps most perplexing of all – do we think about the other plants whose roots share the very same fungal network?[1]

Not all soil is the same. We see huge variability in soils both nationally and globally. From the chalk soils of the South Downs formed from the shells and skeletons of sea animals to the silt soils of Lincolnshire formed from river deposits to the deep peats of the East Anglian Fens, it is estimated that there are over 700 different types of soil in the UK, reflecting differences in underlying geology, topography of the land and climatic conditions.

Soil provides a variety of important functions that can be termed 'ecosystem services' including: helping plants grow; controlling the flow and quality of water; storing carbon; regulating our climate; and providing a habitat for a huge range of different organisms. Importantly, soil grows the food we eat, hosting and feeding plants whose roots grow into the soil, with their stems, leaves and other parts growing above ground fed by air, water and sunlight.

More recently, another factor is having an increasingly significant impact on our soils – us. Human activity is now threatening the ability

of our soils to deliver the ecosystem services on which we rely. Over the last century, especially since the Second World War, the prevalence of industrial farming methods, including increased cultivation, use of synthetic fertilisers and other chemicals, monocropping, as well as significant expansion of the land area under cultivation have all played a significant role in the decline of the UK's soils. The onset of climate change has made soil even more vulnerable, with increasingly hotter summers, wetter winters, alongside expanding infrastructural and property development on green belt land.

What is healthy soil?

As we have started to recognise the scale and threat of soil degradation, the concept of 'soil health' has entered public and political consciousness. The UN Food and Agriculture Organization (FAO), for example, define soil health as 'the capacity of soil to function as a living system, with ecosystem and land-use boundaries, to sustain plant and animal productivity, maintain or enhance water and air quality, and promote plant and animal health'.

Soil is a farm's most important asset, with a healthy soil being the foundation of a resilient farm business. Many farmers recognise that healthy soils are essential to helping them withstand shocks from the climatic extremes of drought and flood. In soil with a good structure, roots can go deep and access the water they need within the soil profile. By tapping into these deep-water reserves, plants can continue to grow through periods of drought when otherwise they might have died. At the other extreme, in times of flooding, a healthy soil profile can store huge volumes of water and can delay the movement of that water into watercourses reducing the risk of flooding and benefitting rural and urban communities.

Soils under threat

In 2020, the UN released a report stating that soil degradation and loss of biodiversity is a major concern for the future sustainability of the planet.[2] In the UK, the main soil degradation processes of concern are compaction, erosion and loss of soil organic matter (or carbon), which has been estimated to cost between £0.9 and £1.4 billion a year in England and Wales.[3]

Compaction and its main causes

When we look at the global status of soils, the figures look catastrophic. In the UK, we have about 13 million ha of land and just over 70% of that is agricultural. And as far as the soils are concerned, about 4 million ha are at risk of compaction.[4]

The physical structure of a soil refers to the arrangement of soil particles and aggregates around which roots grow and air and water move (Figure 3.1). Soil compaction occurs when soil particles are compressed, reducing the spaces between them. Deep or repeat cultivations carried out when the soil is wet will cause soil compaction.

With arable rotations and vegetable rotations, some growers are reducing the depth and frequency of their cultivationto try to protect and improve their soil structure. Growers that use cover crops ensure the soil is never bare – feeding the soil's biology and leaving dead plant material on the surface. That material can be dragged down by the worms for food

SOIL DEGRADATION HAS BEEN ESTIMATED TO COST BETWEEN £0.9 AND £1.4 BILLION A YEAR IN ENGLAND AND WALES.

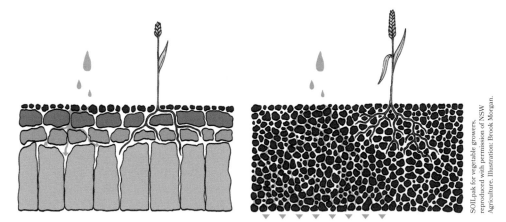

3.1 (Left) Poorly structured and (right) well structured soils.

SOILpak for vegetable growers, reproduced with permission of NSW Agriculture. Illustration: Brook Morgan.

and to generate new, healthy soil, lower down. The fungi, the bacteria, all the invertebrates and worms are all working to keep our soils alive and make them work to reduce flooding, to store carbon and to detoxify the environment. Plant cover really works to support all that life below ground and enhances soil structure.

Nicole Masters describes the number of earthworms that should be found in healthy soil in *For the Love of Soil*: 'an average soil needs to contain at least 25 worms per shovel, 2.5 million/ha. It is not uncommon to find counts less than 10 worms per square spade on chemically managed and cultivated farms, whereas well managed landscapes may have over 70. Numbers will depend on temperature, moisture levels and soil type'.[5]

A healthy, well-structured soil looks like chocolate cake crumbs, whereas compacted, unhealthy soil looks like a concrete slab. That's what happens when the chocolate cake crumbs have their air squeezed out through compaction – it's essentially being suffocated. As the soil is compacted, soil biological activity and root growth are in turn reduced.

Soil compaction also increases the risk of soil erosion by water, as compacted soil is unable to absorb heavy rainfall which means that when we do get heavy rainfall, we are more likely to see water run-off over the surface into our rivers or roads.

Wet soils are more vulnerable to compaction than dry soils as they are less able to hold the weight of machinery or livestock. This is why a large combine harvester can drive over a dry field in August and cause minimal rutting (creation of surface depressions in the soil), but a late potato harvest on wet soils in November can create significant rutting and compaction issues.

Problems of soil erosion and flooding

Over 2 million ha of the UK's soils are at risk of erosion. Erosion can be caused by wind or rain. Water erosion from heavy rainfall is a more significant and widespread problem than wind erosion in the UK. Soil erosion by water can occur wherever soils, particularly those with a high sand or silt content, are exposed to heavy or prolonged rainfall. Any soil condition that encourages surface run-off brings with it the risk of soil wash and erosion.

We have already noted that soil compaction reduces the soil infiltration rate and so can increase risk of overland flow and soil erosion. Topography also plays an important role – large fields with long slopes can accumulate large volumes of run-off water. The Environment Agency has to do a lot of work to reduce flooding and to identify areas at high risk. (See Chapter 5, page 150, for discussion of agroforestry rows controlling waterflow.)

Clearly, erosion risks are greatest where the soil is bare or where there is very little plant cover. Where there is a crop or vegetation, it helps bind the soil together, increases the water infiltration rate and slows down any overland flow of water, reducing its power to erode the soil. This is why the worst cases of soil erosion are always seen from bare fields and why keeping a green plant cover is one of the best ways to protect the soil from erosion.

Eroded soil is washed off the land's surface and carried by the rainfall into rivers, and from our rivers, out to our seas, lost forever. Nutrients are bound to the soil particles, and when eroded, the nutrients can enter the rivers and cause a nutrient imbalance that can be catastrophic for wildlife and also smother spawning fish beds. The cost required to clean up this water is significant, with some estimates stating that about 2.6 million tonnes of soil are eroded each year in the UK.

We now have the additional challenge of a more erratic and variable climate, more so than we experienced 30 or 40 years ago, with predicted rises in global temperatures. Rainfall patterns are changing with longer spells of very dry weather and with rainfall compressed into shorter periods of very heavy fall. Each of our farmers discusses the impacts of this on the land they care for in Chapters 4–6.

Soil erosion and flooding are additionally critical reasons why healthy soils are good for business resilience and business productivity. With improved soil, it's then also possible to reduce input costs, as well as support the environment and in so doing, support and feed communities.

The soil carbon store and loss of organic matter

Soil organic matter is made up of bits of dead and decaying plant and animal material. Soil organic matter is around 58% carbon, and as a result soil is a significant and important carbon store. Globally, soils contain around 2,300 billion tonnes of carbon, which is more carbon than is stored in the atmosphere (800 billion tonnes) and all the planet's plant material (700 billion tonnes) combined. Soil organic matter is a key component of soil health – it is perhaps the single best indicator of soil condition. Most soil functions are driven by biological processes that are underpinned by soil organic matter decomposition. It is the food for the soil biology and what gives the soil life.

Loss of soil organic matter is recognised as a major threat to soil health. When the soil is mixed or cultivated, it encourages the breakdown of organic matter in the soil and some of the soil carbon is oxidised and lost to the atmosphere as carbon dioxide. Continuous arable cropping with annual cultivations and little or no inputs of organic materials has reduced soil organic matter content in most agricultural soils (Figure 3.2). As organic matter declines in soils, they become more vulnerable to compaction and erosion and less able to support crop production.

It has been estimated that around 10 billion tonnes of carbon are locked into the UK's soils, which is equivalent to about 80 years of carbon emissions. They are the biggest means of carbon storage. However, what we've done over the modern industrial age is to take carbon out of the ground as fossil fuels (coal, oil, gas) and burn it.

We need to capture that carbon now and put it back in the soil again to rebalance our planet's carbon distribution. This can be done for free by plants. The carbon that's already in the soil needs to stay there; must be kept there through the adoption of suitable practices, keeping the soil covered with as diverse a range of plants as possible and integrating livestock where soil disturbance can be reduced. These simple techniques can be applied in many ways, and that is what regenerative farmers are doing: taking carbon and putting it back into our land again.

AROUND 10 BILLION TONNES OF CARBON ARE LOCKED INTO THE UK'S SOILS, WHICH IS EQUIVALENT TO ABOUT 80 YEARS OF CARBON EMISSIONS.

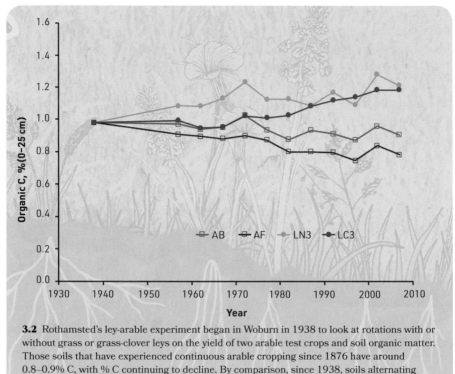

Rothamsted Research, https://doi.org/10.23637/ROTHAMSTED-LONG-TERM-EXPERIMENTS-GUIDE-2018.

3.2 Rothamsted's ley-arable experiment began in Woburn in 1938 to look at rotations with or without grass or grass-clover leys on the yield of two arable test crops and soil organic matter. Those soils that have experienced continuous arable cropping since 1876 have around 0.8–0.9% C, with % C continuing to decline. By comparison, since 1938, soils alternating between 3-year leys and 2-years arable have about 1.2% C. (AB) continuous arable; (AF) continuous arable with root crops or fallows; (LN3) 3-year grazed grass/clover (later grass + N) leys + 2-years arable; (LC3) 3-year lucerne (later grass/clover) leys + 2-years arable.

The sun's energy is captured on land by plants and is ultimately converted to carbon through the process of photosynthesis. The more photosynthesis that takes place in a certain amount of land, the more productive the land becomes. With more carbon in the soil, with healthier soils, the businesses that rely on that very soil naturally become more resilient.

Increasing soil organic matter

Lizzie Sagoo is a principal soil scientist at the agricultural and environmental research and consultancy company ADAS. ADAS was set up by the government in 1946 following the war as part of the Ministry of Agriculture, Fisheries and Food, specifically to provide research and advice for farmers to support food production. The company was privatised in

1997 and continues to provide applied research and development for agriculture and the environment and has a team of soil scientists dedicated to improving our understanding of and protecting our soils. 'One of the most effective ways that we can improve soil health is to build soil organic matter', says Lizzie.

> Many farmers now recognise that conventional farming practices have damaged soil and led to the longer-term decline in soil organic matter and are looking at ways to reverse this trend. We have definitely seen an increased interest from farmers on what they can do to protect and improve their soils – there is an appetite for change, which is fantastic to see.

The soil carbon cycle is a natural cycle. Carbon is added to the soil through plant residues and is lost as carbon dioxide as organic matter is broken down. In an undisturbed state, the inputs and losses of carbon will just about balance. However, in most farming systems, the losses are greater than the inputs and carbon levels in the soil are declining. The good news is that it is possible to reverse that trend by increasing carbon inputs and reducing carbon losses.

> We can increase our inputs by applying manures or by increasing the return of plant residues, for example, by leaving straw in the field rather than baling it or by growing a cover crop or green manure. We can reduce our loss of carbon from the soil by reducing soil disturbance. When the soil is mixed or cultivated, it encourages the breakdown of organic matter in the soil, and some of the soil carbon is oxidised and lost to the atmosphere as carbon dioxide.
>
> We can reduce disturbance of the soil by reducing the intensity of cultivations – so moving from a plough-based cultivation where the top 30 cm of soil is completely turned over, to reduced tillage such as discs or tines which have a shallower depth of cultivation and provide less mixing of the soil, or even to direct drilling, where the seed is placed directly in the ground without any cultivation.
>
> LIZZIE SAGOO

A really good way of recovering soil is by keeping a growing crop in the ground for as long as possible, the green cover protects the soil and the

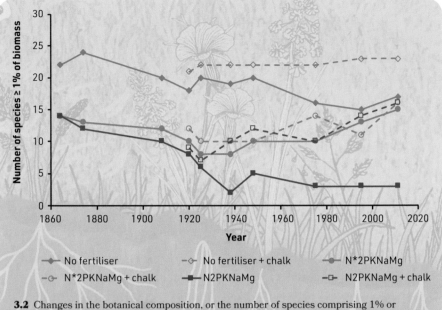

Rothamsted Research, https://doi.org/10.23637/ROTHAMSTED-LONG-TERM-EXPERIMENTS-GUIDE-2018. Fig. 6

3.2 Changes in the botanical composition, or the number of species comprising 1% or more of above-ground biomass undergoing treatments in the 'Park Grass Experiment' carried out at Rothamsted Research. This is indicative of the biodiversity at each site over the past 140 years. The different treatments included: No fertiliser; No fertiliser+chalk; 96 kg N ha^{-1} as sodium nitrate +PKNaMg; 96 kg N ha^{-1} as sodium nitrate +phosphorus/potassium/magnesium/sodium (PKNaMg) +chalk; 96 kg N ha^{-1} as ammonium sulphate +PKNaMg; 96 kg N ha^{-1} as ammonium sulphate +PKNaMg +chalk.

plant root exudates (organic compounds) provide food for a fantastic web of underground microbial diversity. This web of organisms is working to create an open porous soil structure, and to cycle carbon and plant nutrients.

Innovation and Farm Net Zero

Hannah Jones (Farm Carbon Toolkit soil and carbon advisor) says: 'What I get really excited about is when farmers come up with a completely novel situation to a particular challenge where there are opportunities to face that system better, make their farm more resilient and preserve health benefit and biodiversity. As part of a consortium in Cornwall, I work on a lottery-funded community action project where we are building a community of farmers to look at opportunities to reduce greenhouse gas emissions, but at the same time improve farm resilience.'

Cattle farmer Ben Thomas (Chapter 6) at Treveddoe Farm is part of the FCT Farm Net Zero project, and Hannah has been working with him for 2 years. 'I'm excited about what he's doing, including his open-minded approach to integrating trees into his livestock systems. He is increasing biodiversity, developing beautiful soils and meadows with a diverse range of species and using appropriate breeds for his particular environment.'

'What is really interesting about Anna and Andrew Jackson at Pink Pig Farm (Chapter 4) is that they are looking at every single stage and seeing how they could reduce inputs, improve soil health and increase diversity', states Hannah. 'It's then important to try to unpick each factor. Half a field can be enough for an experiment. Then the farmer should assess the field and its performance, and consider what worked, what didn't work and why. Examining, for instance, whether it was down to the success of the cultivation method, or to the species chosen. Regenerating soils and storing carbon in soil is a lifelong process of learning.'

Other great examples of farmer innovation can be found in the Innovative Farmers programme organised by the Soil Association (see page 279).

Tallying carbon credits with soil's variability

The emerging carbon credit market is attracting a lot of attention and needs careful consideration. A carbon credit is a unit of carbon that an area of land has been able to store in addition to normal working activities. The landowner or person who manages the land can potentially sell the carbon that the land will store as a credit or a unit of carbon.

With soils the idea is to follow the reasonably well-established model of woodland carbon capture. With woodland a defined volume and species of tree can be planted on a particular area of land. The plantings are registered and verified. There can be a high confidence level about how much carbon these trees will capture and that they will hold the carbon for a certain amount of time.

We have seen massive increase in interest of soil carbon storage. Farmers and others are starting to recognise the value of that and looking at ways to monetise it. The challenge with soil as a predictable carbon storage medium is that it's highly variable. Its capacity to store carbon depends on the soil type and management practice. Some question the estimates of the amount of additional carbon that can be stored, the ability to measure

it accurately and point out that we have no guarantee that carbon sequestered now won't be lost again in the future. A farmer could get carbon
credits for increasing carbon stored by converting arable land to grassland,
but if the farmer reverts back to arable production in 10 years' time, much
of the increase in carbon will be lost. There is less confidence about per-
manency than with woodland.

Nonetheless there is a lot of hype and several businesses that have already
started selling soil carbon credits. In response, in December 2020, the Sustainable Soils Alliance (see page 281) created minimum standards and a
consultative process to determine a rigorous, verifiable measure of quanti-
fying soil carbon and a recognition of fairness.[6] Having verifiable measures
and fairness across the industry is therefore critical. It must be borne in
mind, however, that experts, like those at Rothamsted Research, have
voiced concerns about overly optimistic views of carbon storage in soils
being presented, with the need for analyses to be given a 'reality-check'.[7]

There is debate about whether selling soil carbon credits is the right
thing to do. Some people recognise the ability of the soil to sequester
carbon and think that if farmers can receive money to support them to
adopt management practices that allow them to sequester carbon, then
this is a good thing. Particularly when there is a cost to adopting associated better land management practices, which the sale of carbon credit
could offset. Others question whether it is ethical to sell carbon credits
which allow other industries to continue to pollute. Dr David Powlson
(formerly at Rothamsted Research) has warned against carbon credits
being treated as a get of jail free card, for example.

Ultimately, organic matter (and soil carbon) is central to soil health,
and therefore, increasing organic matter will have a range of benefits, from
improving soil structure, increasing water holding capacity and increasing
soil biological activity. If the soil carbon market can fund and drive more
beneficial change within a rigorous, fair and verifiable system then that
would be welcomed.

The challenge of measuring carbon

The next question is about how to measure carbon in the soil effectively.
Carbon stored in the soil is in the form of organic material. The sugars
and proteins and all the various biochemical components in that matter
have carbon-based complex molecules. The organic matter breaks down
at different rates. Some forms of organic matter break down quickly and

others more slowly, for example, grass clippings (simple carbon) versus bark chippings (complex carbon). Bacteria, fungi and plants all help break down this complex material into sugars or proteins:

> In a world devoid of its army of decomposers, we would rapidly be overcome by debris and despair. Microscopic bacteria, fungi, protists and insects of every size are the critical clean up crew, digesting organic materials and keeping our world turning. They have an extraordinary ability to break down them into bio-friendly foods for other organisms.[8]

There is very rapid recycling of organic matter, releasing carbon, whereas complex forms of carbon, which are resilient, will stay in the soil for many years, sometimes decades.

FCT measures soil carbon by analysing replicated soil samples using a standard method. Soil samples are sent away to the lab. There are two tests, the first being a loss-on-ignition test measuring the difference (loss) in mass of the sample before and after burning at high temperatures. This provides a measure of the organic matter content (which is lost by burning) and can indicate the quantity of various molecules (including carbon-containing molecules) that were in the sample before burning. The other type of test is the Dumas test, which provides a number representing the total carbon within the soil sample. Either figure can be used to estimate the total soil carbon stocks in a field, which in turn can be converted to an estimate of how much carbon makes up the organic matter in the soil sample. With these estimates, one can then approximate the total carbon within the field. (See Chapter 6, p. 207.)

With the increased focus on soil health, more people have been asking how to measure it. How do we know if our soil is healthy and how do we monitor change over time?

Measuring soil health

The Agriculture and Horticulture Development Board (AHDB) is a statutory levy board funded by farmers to deliver research and other work for the benefit of the farming industry. Soil health was identified by farmers as a research priority area for AHDB, and in response AHDB set up the Soil Biology and Health Partnership in 2015. The partnership was a 5 year programme of research and knowledge exchange delivered by soil scientists from a number of universities and research institutes, including NIAB (National Institute of Agricultural Botany) and ADAS. It was designed to help farmers and growers maintain and improve the productivity of UK agriculture through better understanding of soil biology and soil health. The partnership worked closely with farmers, advisors and the wider industry to draw together and build on knowledge and experience to develop a soil health 'scorecard' – a health check for our soils. The scorecard includes measures of soil physical, chemical and biological health, including soil nutrients, organic matter, soil structure and earthworm numbers. Typical ranges are given from each of the soil health indicators along with a green, amber or red traffic light system for highlighting where any of the values differ from the norm or typical ranges that would be expected in a healthy soil.[9]

The simple 'slake' test for biological activity and soil particle aggregation

The slake test is a simple test that compares the condition of soil from different areas of a farm (Figure 3.3). This test involves collecting and placing soil aggregates (clumps) in a clear cup, jar or other container of water and observing how long the aggregates stay together. Soil microorganisms secrete glues which stick soil particles together. Soil aggregates from a healthy biologically active soil will tend to stick together in water, whereas soil aggregates from a damaged soil will tend to rapidly disperse in water. The slake test gives us a good visual indication of what will happen to soil when hit by heavy rainfall. A soil which rapidly disperses in the slake test will be more vulnerable to soil erosion following heavy rainfall.

3.3 (Left) Microbially healthy (sticky) soil that has kept its integrity as it's passed through the water to sit at the bottom of the glass. (Right) Soil that has been heavily cultivated. On contact with the water it disintegrates.

Soil science on the political and educational agenda

The last 20 years have seen a number of reports from national and international organisations on the threats facing soils and calls to action to protect and restore soil, including from the FAO, European Commission and individual national governments. The UN declared 2015 as the International Year of Soils, and every year the 5 December is celebrated as World Soils Day.

In December 2023, the UK Environment, Food and Rural Affairs Committee (EFRA) published the report to its inquiry into soil health. The committee's report calls for soil health to be put on the same footing as water and air quality within government policy, with statutory targets on soil health, alongside the existing water and air quality targets, by 2028.[10]

Farm Carbon Toolkit

The FCT was developed by farmers for farmers. For over a decade, FCT has worked to further the understanding of greenhouse gas emissions in agriculture. It provides tools and services to measure impact and runs projects with farmers that inspire action on the ground. Its vision is a farming sector that minimises its carbon emissions and maximises its carbon sequestration while producing quality food and a wide range of public goods, all produced by resilient and profitable farm businesses.

FCT supports farmers to measure and understand the greenhouse gas emissions from their land and provide an open discussion of how these greenhouse gases can be reduced.

The Toolkit can be used with a farm carbon calculator, and also runs the Soil Farmer of the Year Award, which finds, promotes and champions farmers who are passionate about their soils, safeguarding soils and improving what they do. FCT also offers:

- **peer-to-peer training and learning events** on whole-farm and soil carbon footprinting
- **farmer outreach and support projects**
- **bespoke advice and action planning with farmers, landowners and companies**, centred around measuring and

Farm Carbon Toolkit *continued*

improving soil health, greenhouse gas emissions and other ecosystem services

- **partnerships with research institutions** to further the understanding of soil carbon sequestration and best practice protocols for measurement, reporting and verification.

Some individuals are doing a great job and take risks on their farms – it is these farmers that we are keen to celebrate. They take up the challenge to see if they can make something work even better. If one celebrates what's good, there's an opportunity to improve further.

Hannah Jones (FCT soil and carbon advisor), 'I do what I do because I belong in fields, digging and looking at soil. I belong sitting around a farm kitchen table, working with farmers to understand their farm systems and the challenges they face. I come from a farm background. I know that farmers are dealing with many different concurrent things, and I'm aware initially we may be an unwanted additional draw on their time. I've worked with some really inspiring people in the past at universities, at charities and many other farmers who have all inspired me to do what I do now. And it almost feels that this is really where I belong.'

The British Society of Soil Science has been working to raise the profile of soils in education and has supported the development of a new Level 7 Soil Science Apprenticeship. At the same time The Country Trust's (see page 271) 'Plant Your Pants' campaign has been taking soil science into primary school classrooms.

Plant Your Pants

New survey data from The Country Trust shows that nearly a third (27%) of children between the ages of 7 and 10 years old have never, or hardly ever, got their hands dirty in the soil. Additionally, half (50%) of children surveyed simply see soil as 'just dirt'. The Country Trust aimed to change that with its nationwide citizen science campaign, launched in 2022, called 'Plant Your Pants'.

Pants that are made from cotton are organic matter. The living components of the soil, such as earthworms, nematodes, springtails and microbes will help break down cotton pants.

A healthy soil will show significant breakdown of the pants in a given time period (8 weeks with Plant Your Pants). If the pants don't break down significantly, then there might be fewer organisms alive in that patch of soil, or the soil might be too damp or dry to act on the pants. When it's too damp, organisms struggle to get the oxygen they need in the soil when all the spaces in the soil are filled with water.

Plant Your Pants experiments take place annually between March and June, with results being uploaded on The Country Trust's interactive soil health map once complete.

Soil further reading

The UK Centre for Ecology & Hydrology's short article, 'What is soil?' (https://cosmos.ceh.ac.uk/soil) provides some accessible further detail. For those wanting to dig a little deeper, Rothamsted Research's Repository, https://repository.rothamsted.ac.uk, contains details of the institution's research output from 1843 to the present day. Nicole Masters' *For the Love of Soil: Strategies to Regenerate Our Food Production Systems* is an excellent resource.

Interlude II

Regulation, greenwashing and co-option

VICKI HIRD

Regenerative agriculture (aka regen ag, regen and RA) is a relatively new term describing what may be a sustainable farming system. Chapter 2 provides a full description of its characteristics and where it sits in the wider agroecological approach. The concept has been around since the 1980s but has recently seen a huge rise in popularity and attention.[1] It's also been recognised by the agri-food industry as a possible solution to certain issues, like carbon emissions, and as a marketing opportunity.

To regulate or not to regulate?

Regenerative agriculture lacks a formal, independently set definition, and that causes problems. It can mean different things depending on who is using it and in what context. Advocates would agree that at its

core it should involve some mix of the following farming practices:[2]

- limiting soil disturbance
- maintaining soil cover
- fostering agricultural diversity and rotations
- keeping living roots in the soil
- integrating livestock and arable systems.

Many regenerative farmers will go further. However, not all regenerative farmers, or the corporations that promote them, use all these regenerative tools, and their use does not automatically indicate a good farm system overall or guarantee nature protection or recovery.

Freedom from a formal definition or regulation can also be beneficial. Changing a farm system comes with considerable risk. For many farmers who are struggling just to survive, whole-system change may simply be beyond them, but the flexible regenerative approach encourages farmers to take the first step on what many refer to as the 'regenerative journey'. They can try out one of the tools, assess the results, adjust, and move on to another. Many of the case-study farmers in the film and this book promote the idea of giving it a go, of starting small and building from there.

A good example of a farmer-led nascent standard is Wildfarmed, who are growing in their impact and reach.[3] They call themselves the 'gold standard' for regen and have managed to interest big buyers. They are third party audited and their website states:

Regenerating our soils, health, water, wildlife and food requires a complete mindset shift from control to nurture, linear to circular, extractive to regenerative. It requires community and a new kind of supply chain.

If that last part works, it could be a valuable whole-system approach and, as the farmers in Wildfarmed include organic and conventional and large and small systems, there is potential to achieve cohesive environmental and social outcomes.

In contrast, the large-scale, corporate-led 'regenerative' farming programmes (and big associated advertising campaigns), while looking good on paper, may fail to deliver long-term outcomes – given the imperative

to drive up yields, keep raw-material costs low, and secure profits in a highly competitive, increasingly processed food system.

The increased interest in regenerative approaches is bringing investment in both farms and research. Long-term experiments on regenerative approaches, such as no-till and diversified cropping, show that they are not a short-term fix or alternative for more holistic, sustainable food production systems.[4]

Greenwashing accusations

Co-option of regen ag by corporations and prominent members of the food and farming industry is of concern. In the corporate rush to reduce the environmental or carbon footprint of their food supplies, large companies are making big claims and placing demands on farmers, without offering financial support to achieve them.

There has been a huge amount of hype recently around the ability of regenerative approaches to sequester carbon, and companies are using this to claim major emission reductions for their products and their supply chains. This has led to considerable accusations of greenwashing. There are huge uncertainties around the claims of soil's high carbon content, and more independent research, unconnected with industry funding, is needed to address this. The adoption of no-tillage, a common and core regen approach, does increase soil organic carbon content in the first 100 mm of soil, but a meta-analysis of the research suggests that the gain may be undermined by reductions in carbon content at 100–400 mm depths. This could mean there is no overall change in carbon content, so measurement needs to include the whole system, and carbon-offsetting claims need to be seriously challenged.[5]

A lack of a legal and independently verified definition for regenerative agriculture is not helpful in this regard. Nor is the lack of clear impact. Often claims are made to the detriment of existing, verified systems, such as organic – setting regenerative agriculture in opposition to organic systems, especially over the use of ploughing/tillage. No-till agriculture is extolled as the way forward, thereby vilifying the use of ploughing in organic systems, which is a key tool for weed suppression. No-till systems frequently rely on herbicides, which may be more harmful than tilling (see discussions in Chapter 4).

To avoid misleading claims, the key will be to have independent standards and clear outcome-based auditing on the impact of specific agricultural practices. PepsiCo, for instance, has a huge global supply chain, including corn, wheat, oats, soy, cocoa, potatoes and orange juice. They have committed to spreading regenerative practices across their key-ingredient footprint, equal to 2.8 million hectares.[6] They estimate that regenerative farming will cut at least 3 million tonnes of their greenhouse gas emissions by the end of this decade. These are fine words, but it's worth noting their 'rules' state: 'PepsiCo does not require fully [sic] traceability or segregation to the farmer or farm land for a regen ag or climate claim'.[7] Some have accused the company of greenwashing, given their overall role in globally driving intensive farm systems, monocultures, low farmgate prices, the loss of small farms, unsustainable water use, excess plastic use and unhealthy food impacts.[8]

The 2023 FAIRR report, 'The Four Labours of Regenerative Agriculture',[9] assessed the regenerative agriculture commitments of 79 large, publicly listed agri-food companies, collectively worth US$3 trillion, to see if they were meaningful. These are some of their findings.

- Of the 79 companies, 50 mentioned regenerative agriculture initiatives in their disclosures.
- Of those 50, only 36% (18/50) had quantified company-wide targets for regenerative agriculture.
- Just 16% (8/50) discussed metrics and data, with only four companies having established baselines to measure progress.
- Only 8% (4/50) had targets to financially support farmers to deploy regenerative practices.

All the major global food corporations, including General Mills, Nestlé, Danone, Coca-Cola, and the big commodity traders, including Cargill, promote their activities on regenerative agriculture. Is this a serious attempt to reduce the huge impact of the intensive farm systems they rely on for cheap food ingredients? Maybe some of it is, but the huge marketing efforts being employed, the claims being made and the lack of clear benefits to farmers and society are certainly under question.[10]

What is needed

There is a clear imperative to have independent, verifiable standards for regen without undermining the need for continuous review and improvement, and with strong recognition of the needs of farmers and growers. Any accreditation scheme must spread the risks and benefits so farmers do not take all the risks of changing their system and bear the expense of reporting and accreditation while gaining no reward. Transparency on standards, accreditation, impacts, fair dealing and sourcing would ensure all those in the supply chain involved with this better product get paid a fair price and that the scheme standards are clear to everyone, including the consumer.

Government regulation of related carbon, biodiversity and other offset markets associated with any farm scheme is vital to remove abuse, provide clarity for offset buyers and ensure a level playing field.

More research is clearly needed on what is possible and verifiable. As researchers looking at the available evidence on regen outcomes have noted, 'Despite the purported benefits of RA, a vast majority of growers are reluctant to adopt these practices due to a lack of empirical evidence on the claimed benefits and profitability.'[11] The same study suggests there can be major variation among different agro-ecological systems and recommends implementation of 'rigorous long-term farming system trials to compare conventional and RA practices, in order to build knowledge on the benefits and mechanisms associated with RA on regional scales'.

Greenwashing needs to be challenged from a whole-system perspective. Most purchases are currently from big retailers, food manufacturers and food-service companies. We need to be sure, if they are promoting 'regen', that this means whole-system approaches that genuinely deliver for nature restoration and climate resilience, alongside the healthy food we need.

4 Regenerative mixed farm

Pig Pink Farm, Scunthorpe, Lincolnshire

ANNA & ANDREW JACKSON

THIS IS OUR PATCH OF LAND. WE WANT
TO DO EVERYTHING IN OUR CONTROL
TO BE AS ENVIRONMENTALLY FRIENDLY
AS POSSIBLE. WE'LL DO IT WITHOUT
EXTRA PAYMENT.

IT'S FOR EVERYONE, FARMING.

EVERYONE EATS, EVERYONE LIVES ON
THIS PLANET.

FARM PROFILE

YOUR NAME(S)

Anna Jackson and Andrew Jackson

REGION

North Lincolnshire

SIZE *(land area owned/managed)*

324 ha over 2 sites

RAINFALL *(average annual)*

710 mm

SOIL

Sandy to clay-loam; calcareous

TENURE
(of land ownership)

Owned

KEY AGROECOLOGICAL PRACTICES

No-till, regenerative arable cropping, with integrated sheep

BUSINESS/FARM NAME

Pink Pig Farm

BUSINESS TYPE
(e.g. mixed farm, smallholding, forestry)

Mixed farm

PRODUCTS *(principle business outputs)*

Wheat, oilseed rape, grass seed

ALTITUDE

20 m

EMPLOYEES *(include yourselves)*

3

YOUR APPROACH
(self-defined, as you see it)

Regenerative

WEBSITE

https://www.pinkpigfarm.co.uk

Dad and I farm across two sites near Scunthorpe in north Lincolnshire, the second biggest county in the UK. We're also close to Brigg, an agricultural market town that sits between our two holdings, which Dad says has everything a farmer could need, including a farmers' market that we've attended as a trader for 20 years. In total we farm about 800 acres (324 ha), with around 750 acres (304 ha) cropped. We've got some woodland, bits of an old aerodrome and some permanent pasture accounting for the rest. The farm size, for this region, is about average although probably larger than the average in other parts of the country.

The two farms are about 13 miles (21 km) apart. Pink Pig Farm, where we live, has the Pink Pig Farm attraction and restaurant, our main diversification. My great-grandfather came here in 1929. It's 300 acres (121.4 ha), half sandy land which is not very productive, and half better-bodied clay-loam. The second farm is 450 acres (182 ha) of uniform, free-draining calcareous soil, which Dad says is idiot proof, but we can still seemingly make a mess of things sometimes.

We grow wheat, our most stable crop; vining peas, which we've grown for about 40 years; and oilseed rape, which despite being a bit more of a high-risk crop is widely used by farmers between wheat crops. Recently we've tried other things with varying degrees of success: lupines, grass seed, quinoa and meadowfoam, which produces a cosmetic oil. Some farmers, and it's their choice, grow wheat, rape, wheat, rape, wheat, wheat, rape, rape. Whereas, we want to have as broad a rotation as possible, which is really good for the soil, good for the wildlife and good for diversification. That's our plan. We've tried different break crops in between the wheat, and some haven't paid their way. We continue to experiment. We're introducing the fifth element of regenerative agriculture – livestock on arable land (see Chapter 2) – with an expansion and integration of our sheep.

Why we farm

ANNA: I wasn't going to come into farming at all. I looked to the conventional system that we were caught up in, and it didn't inspire me. I'm definitely an environmentalist at heart; I love our planet. There was nothing – no offence, Dad – that dad was doing that was inspiring.

DAD (ANDREW): Although I grew up on the farm, I didn't know much because my father hadn't really taught my brother and I about it. My fondest memories are of the summer holidays when all my friends came up, and we just played around on the farm in tree and bale dens. If we did any farming, it was just riding in tractors and trailers at harvest time. When I was 15 or 16, I might have done a little bit of tractor driving at harvest to earn a bit of money.

I left school in about 1975. I went into an agricultural college's tent at the local show and thought, 'Well, this beats doing A-levels.' Prior to the college course I attended day release, and I learned a lot while out on farm. I came back from college to the farm in about 1980. Everything about farming I just love. Every day is different. You have to be a jack of all trades, master of none, which is just about what I am. It fitted and I'm still here.

ANNA: When I was younger, Dad said, 'Don't come back to the farm, there's no money in farming, go away, get an education, do something else.' So I did. I became a freelance sports photographer and was living in London. I also co-ran a project, with a friend, called Zero Waste Life. We both tried to live zero waste for a year, to personally create no rubbish. And we did it. The waste fitted in a single jam jar. It really taught us so much about what we're doing to the planet.

Then COVID hit. All my clients went silent. I had no way of making money, and I thought, 'I've got to go home, really, because I can't pay rent and I need a roof over my head.' Mum and Dad were fine with that, so I came home at the age of 28. I got bored quite quickly, so Dad gave me an orphaned lamb and asked me to try to keep it alive. I did – Timmy survived and he's still on the farm today: a huge, castrated male sheep. Timmy was probably the catalyst. He showed me how great farming can be, how great sheep are and how fun it can be to work the land.

Dad had given me a book, about 3 months before the COVID outbreak. I read it, which was quite unusual because it was a farming book, and I didn't read farming books. The book was *Dirt*

I FOUND DAD HAD GONE REGEN SORT OF SECRETLY, HE HADN'T SHOUTED ABOUT IT, HE JUST STARTED MAKING CHANGES.

to Soil by Gabe Brown. When I came home, I found Dad had gone regen sort of secretly, he hadn't shouted about it, he just started making changes. Here were the principles from the book being introduced on the farm, and I started to see.

DAD: That book is probably the best book I've read. It's been a turning point for a lot of people. Once I'd read it, and realised this guy's been in it for 30 years and it works for him, I thought why not try it?

ANNA: Dad showed me everything he was doing on the farm, how he'd stopped disturbing the soil, how he kept the ground covered at all times, how he had introduced sheep into the rotation, how he was using diverse cropping – all of this cool stuff which helps the environment and also has increased his profit slightly. It really lit a spark in me. I love the outdoors, I love sheep and I love the environment. And it all came together in farming. So I said, 'Shall I stick around?' And you were like, 'Yeah, all right.'

If Dad had been farming conventionally, like he was when we grew up: using all the chemicals, ploughing up the land and disturbing all these habitats, then I wouldn't have gone into farming, for sure. Now he's farming with a conscience, he's farming with a purpose. Dad said he wants to improve this farm as much as he can before he dies, and because of that, it's a very different farm.

DAD: Most farmers, including myself, put into action what they're taught at university or agricultural college, which was best practice at the time. We were doing what was asked of us and were just trying to make enough money to survive.

I'm quite an environmentalist myself; for about 10 years we had an organic certification, but that wasn't quite for me. I felt it was a bit restrictive and found it difficult to make money, so we came out of that. But once you've had a change of mindset, you don't go back. We evolved into this regenerative practice because I've got an inquiring mind, and I wanted a better way. I couldn't argue with anything in the books – everything seemed to stack up. I want to try to put it into practice and achieve what Gabe Brown did in 30 years, but in fewer, in 10 or certainly before I retire or die. We're in a lucky position being third generation on the land and owning it, so we can probably afford to take some risks that other people wouldn't or couldn't, especially if they're on rented land.

Essentially, we've become a sort of experimental farm. We're trying things to prove they will work. Other farmers around the country are doing similar, and if we can prove it works, that it's good for the planet and the environment as well, then hopefully more will take it up.

Two jobs

The team is me, Dad and Carl. Carl and I are both part time. Carl has his own business as a contract sprayer. I have two jobs at the moment because there's not enough money in farming for me to be able to just farm. I work for my mum on the social media for the Pink Pig Farm attraction. Luckily, they run well alongside each other. I would like to just be a farmer, but I'm not sure that will ever happen.

I need to find a way financially that I can farm full time. Perhaps we could rent more land, so we have a bigger income, or we look for more diversification on the land we have. I've got to be resourceful and find a way to keep farming because this is what I love doing. This farm supports two people and at the moment there are three of us.

It's ridiculous when you think about it. We are a typical farm and we don't produce enough money to support a small family. People ask if that makes me angry, and weirdly it doesn't because, coming from a background where I had my own business, I know what it's like to be self-supporting – I know what it's like to have to hustle for yourself. A lot of farmers have been receiving the Basic Payment Scheme (BPS) subsidy. When I found that out, as a self-employed business owner, I was shocked, I thought, 'The government doesn't give me money to do my job.' I was a photographer. I wasn't providing food for the nation, let's just get that straight, but still I think it's kind of remarkable. It blew my mind that it was happening at all, that there was a need for a subsidy. Why aren't food costs higher? You know, people paying more for good produce rather than the government supplementing all the farmers. It just seems like a broken system.

A testing time

Growing up, we didn't really get to see our parents that much. They were doing so much to try to make the farm profitable. It wasn't enough to just farm the land. Mum set up the Pink Pig to bring in an additional income source, and we added the farm shop. The diversifications brought more income but more pressure too. We got to see them as we were living on the same farm, of course, but their brains were constantly elsewhere. That was probably another reason that pushed me away from wanting to be involved in the farm.

Dad says the organic period was very complex. We were trying to keep pigs and grow parsnips, carrots and potatoes, and have laying hens. Then we started growing products for the farm shop. So another 10 additional products on top of Dad's day job running the normal farm. He was keeping a lot of balls in the air.

Dad set out to supply organic produce to the supermarkets. He saw it as a niche in the market that might bring higher returns, and it fitted with his green leanings. The marketplace is very small, and if you get a product rejected for any reason, there's no outlet. Every product you send to a supermarket has to hit their specification. It doesn't matter whether it's potatoes, pork or parsnips. One year the potatoes got black scurf, which is a fungal disease affecting the appearance of the skin. When you've got organic potatoes that don't meet the quality criteria, nobody else wants them. With the pork, when the first batch of pigs were ready to go, the supermarkets added another quality-assurance hoop for us to jump through. We spent 6 weeks getting this extra quality assurance. By that time, the pigs had eaten their way through the profit and gone out of spec, so we got penalised on the contract a lot. The parsnips came in undersized due to weed competition, and a lot were rejected, which is ironic now as they charge extra for 'baby' ones and make a big noise about being all green by selling 'wonky' veg.

We won't work with supermarkets anymore because they're just awful to farmers. And I feel really sorry for the farmers that do because you can't negotiate. There's no way that you can tell them how much your produce is worth. You just have to go with whatever price they offer. And if they say, 'Oh, you know, we don't want that anymore', then you've got to find another way to sell it at the last minute before it all goes off.[1]

It was a bumpy ride, but one that evolved into a farm shop where we're in control – selling our own organic produce direct to the customer alongside the Pink Pig attraction. Unfortunately, we don't have the shop anymore. We found sending somebody out to pull carrots each day, then wash them, grade them, cool them and present them, didn't really make commercial sense. In the end we were buying quite a few products in – and that's not the ethos of a farm shop. It's not that we don't want a farm shop, it's just that we didn't have enough people wanting to buy from it. We were trying to provide people with local food, organic, home-grown, like literally grown here, sold here, but people in this area, possibly because of the economy, just didn't want to buy it.

Becoming regenerative has freed up a whole lot of time and saved a lot of money in wearing out equipment, replacing machinery and diesel expenditure. I think now, compared to when we were younger, it's a much healthier lifestyle. Yes, we're still very involved. We work particularly hard during lambing, the harvest and drilling; we work all the time the sun is up (and around the clock with lambing). But it's focused on 2 or 3 months of the year, and in other months, we actually have some slack for a bit of a life. It's better for mental health; I'm able to see friends and keep friends, which is important.

WE WON'T WORK WITH SUPERMARKETS ANYMORE
BECAUSE THEY'RE JUST AWFUL TO FARMERS.

Not bigger, different

In order to stay profitable, a lot of farms get bigger, seeking efficiencies, economies of scale, chasing bigger yields on smaller margins, taking them further down the industrial road. Dad chose not to get bigger but to get different.

DAD: It's probably fair to say it started with Twitter. I started looking at Twitter because my son told me to get on there. There were farmers detailing the changes they were making. I was pointed towards a couple of books: David R. Montgomery's *Growing a Revolution* and Gabe Brown's *Dirt to Soil*. I saw that this was a good-news story. It was a pretty tall order, but it was something to work towards. And so I decided that I would take that leap of faith. It's a long journey; Anna's on board with the concept and we're working towards it together.

ANNA: It's going to be a slow process, but it's worth it. The reward that you get from it is immense.

DAD: I would like to have a farm that's very diverse in wildlife, and I know we can improve our soils dramatically. I want to leave a better farm than I inherited. This is going to take time. Nobody likes sprays or chemicals, but we can't just go cold turkey. We've got to do it gradually. We've taken on a regen consultant to help us, and our existing agronomist is also on board with what we're doing.

I've joined BASE-UK (biodiversity, agriculture, soil and environment), which is a group of farmers who share knowledge on regenerative farming (see page 270). Membership in the last 2 or 3 years has grown from 200 maybe to over 500 now. We have Zoom meetings, an annual conference and farm walks to promote regenerative and conservation agriculture. It's very supportive and I've gained a lot of knowledge from other farmers.

ANNA: If you're nervous about making changes, just start slowly. You don't have to go into all the mycorrhizal fungi and all the long, complicated words that appear when you mention regen. You can just reduce your movement of the soil and your movement over it, get some livestock on your fields and start slowly reducing your chemical inputs and replacing them with natural fertilisers.

DAD: At heart it is a very simple concept: mimicking nature. If it were complicated, I couldn't do it.

ANNA: I think people sometimes think it's super complex, and that they're not going to be able to do it because they don't understand all the science jargon. We didn't take any courses or anything. Dad just started.

DAD: The biggest hurdle is changing your thinking.

Looking to the past in a positive way

Some people say regenerative farming looks to the past; this is sometimes levelled as a criticism. It does look back but in a positive way. We've talked to an old hand, Reg, who's worked on this farm all of his life and has basically seen the whole industrial revolution in farming happen. He's looked at what we're doing now and said that some of it is what he used to do. Dad found details of a rotation in our old archives, from my great-grandfather's time, showing they were planting a very varied rotation back then.

In regenerative agriculture, rediscovering older good-practice is being integrated with new approaches and techniques. Take for example the use of the herbicide, glyphosate, regenerative farming, in the main, still uses it. This can particularly be the case in a no-till system, to terminate a cover crop, and there is a lot of discussion of how we might stop. Gabe Brown in the US and Tim Parton here in the UK dispense with or greatly reduce glyphosate use by roller crimping cover crops when they are subjected to a heavy frost, with great results. Of course, in parts of the country without reliably cold frosts this isn't an option.

Looking to the past, we rogue our wheat fields. We walk up and down the tramlines, looking left and right, scanning as we go. When we spot something that shouldn't be there, we pull it out, knock the soil of the roots, and put it in a big rogueing bag which we carry on the shoulder. When growing wheat, we are looking out for volunteer crop plants like barley or oats, and weeds such as black grass and brome. The seedbank, Dad says, can hold seeds dormant for a hundred years, so anything can pop up.

IF YOU'RE NERVOUS ABOUT MAKING CHANGES, JUST START SLOWLY. THE BIGGEST HURDLE IS CHANGING YOUR THINKING.

Glyphosate

The herbicide, glyphosate, marketed under the brand name of Roundup, has been available since the mid-1970s. It is a broad-spectrum systemic herbicide and crop desiccant. It was widely adopted for agricultural weed control, especially after the agrochemical industry developed glyphosate-resistant crop strains, enabling farmers to drench their fields in it, killing weeds, killing everything but the crop. It is also used as a pre-harvest desiccant on non-resistant crops: killing the plants and promoting drying before harvesting. Initially expensive, its price dropped considerably when it came out of patent in 2000.

It is an organophosphorus compound, specifically a phosphonate. It was first synthesised in Switzerland in 1950 and patented in the 1960s, not as a weedkiller but as a chemical chelator (to bind and remove metal particles). The US agrochemical company Monsanto was working with it as a potential water-softening agent when its herbicidal activity was recognised.

There is ongoing scientific debate over its role as a carcinogen. The US Environmental Protection Agency consider it to be 'not likely to be carcinogenic to humans'. Whereas the International Agency for Research on Cancer (IARC) has classified glyphosate as 'probably carcinogenic to humans'.[2]

What we're growing, wheat (left), and what we're pulling out, barley (right)]

Dad has always done it. When I was little, he would ask, 'Does anybody want to come rogueing?' And we would obviously say, 'No!' At the time, when farmers started putting herbicides on the fields to kill off all the weeds, Dad was not so sure. So he decided to keep the practice of rogueing the fields. We hardly have any black grass because we've kept it up. Some fields we don't have to spray herbicides on at all because our rogueing is so effective, and in others we may just spray a localised bad patch.

We are literally walking the whole field up and down, up and down. As you're walking, you're observing, spotting disease in the crop or where a patch has become flooded and boggy. You'll notice if there are more birds or insects – you'll see what's going on with your land. In contrast, when you just drive over a field, spraying, you're not really seeing, but when you walk, you're really in tune with what's actually going on in your fields. We're not a big farm so it is manageable. This land is our livelihood; it's important to us and we should give it time and really inspect it. It took me about 6 months to get the hang of it. Now I'm tuned in and I can't help myself. I'll see a weed and have to tell myself, 'Anna, you don't need to pick that. You're in the park!'

Another older technique we're using in our oilseed rape fields, instead of desiccating with glyphosate, is swathing. We come in with a swathing machine which cuts the bottom of the stalks so the rape falls into a swath, resting off the ground on the stubble. Once it's cut, it dies and begins to dry out. It has the same effect as you would get with glyphosate, but instead of chemical desiccation, it's a natural process. Then you leave it for around 10 days before harvesting. A further benefit is that we retain our companion crop of clovers, a chemical desiccant would kill them off. So far, it appears to work quite well.

There're a few reasons why people stopped swathing. One of them was weather unpredictability; once you've swathed, your crops are laid out precariously, and if there's something like a big hailstorm or rainstorm, you could lose your crop. But the main reason was glyphosate came along and made it easier.

THIS LAND IS OUR LIVELIHOOD; IT'S IMPORTANT TO US AND WE SHOULD GIVE IT TIME AND REALLY INSPECT IT.

No-till

One of the key principles of regenerative agriculture, usually listed first, is to minimise soil disturbance (both chemical and mechanical) by minimising or stopping tillage (ploughing) (see Chapter 2).

ANNA: When Dad found out about no-till, he dramatically sold the plough.

DAD: Well, no. I stopped using it. Then when we hadn't used it in 5 years, and I thought it might lose value, I sold it.

In the 1970s there were some people experimenting with no-till, which means sowing the seed directly into the soil without inverting it with a plough or breaking the surface with the cultivator. But it didn't catch on here, even though the Americans were actually making it work. It wasn't until 20 or 30 years later that farmers in the UK started to try what the North and South Americans were already doing.

It seemed to me that you were mimicking nature better by doing this and there were multiple benefits: retaining and building up soil organic matter, not bringing weed seeds up to the surface (so eventually your weed burden may go down) and using less fuel, wearing out machines less, spending less time in a tractor and compacting the soil less. Everything seemed right about it. So I bought a strip-till drill, which only cultivates the portion of the field that is planted, leaving undisturbed strips either side. We ran that for 3 or 4 years. It was a great drill for levelling the fields and was a steppingstone to the no-till drill that we've got now, which just cuts a slot with a disc, and then the seeds go in the slot and the slot is closed back over. We're doing it and it works.

ANNA: We want minimal soil disturbance to keep the biology underneath the soil as happy as possible, and the no-till drill gives the least disturbance.

DAD: Cultivation destroys soil networks – the mycorrhizal fungi that help the plants reach out and collect the nutrients from the soil. Whenever you move the soil, you excite the microbes, and they eat the organic matter, emitting carbon dioxide into the air. We've been measuring our soil for carbon content, measured principally in organic matter, and we've identified that the soil in the hedgerows, which is a good indicator of where we need to be, is at 6%

Soil comparison: (left) mid-field and (right) bottom of hedgerow.

or 7%. Our soil in the fields was down to 2% or below in 2012. We're now averaging about 3.6%. I'd like to carry on building this back up to 6% plus.

The role of sheep on our farm

Dad's always had sheep, but a small flock, more of a hobby. We've been building the flock up from the 60 Romneys/Lleyn we had in 2021 because we want them to graze our cereals and cover crops, to introduce natural fertiliser back onto the land and take their place in the new rotational system. It's been 80 years since animals have been on our crop land, which is just peculiar, given that livestock had so central a role on mixed farms in the past. Looking at my great-grandfather's invoices, we can see there were good livestock numbers. That stopped in the 1940s and 1950s when we started buying artificial fertiliser. Sheep are incredible because they're going to poo on the land, trample it in, and munch the top off the crop, which forces it to shoot more vigorously. Being lighter than cattle, there's also less of a risk of poaching the ground in the winter.

We use a mob-grazing system in which we move the sheep often, so they don't overgraze an area of land. It's also good for them because they get fresh pasture, and there's less of a parasitic worm burden. People use different intervals, and it depends a lot on your land. I like to move them every 2 to 3 days. They love moving; I'll call them, and they come running.

We trialled a no-fence system. The sheep each wore a GPS collar, and I set a digital parameter on my phone. It was good as I could track where they were and move them on more easily, but because the system was quite expensive, given the return on the sheep, we've had to stop using it and have gone back to temporary fencing; I think it'll work better with cattle, financially.

ANIMALS HAVE THE NATURAL INSTINCT TO EAT
WHAT THEY'RE DEFICIENT IN, TO HEAL THEMSELVES.

I've introduced a mineral bar, which was suggested to us by our regen consultant, Ben Taylor-Davies. I've been providing molasses, apple cider vinegar, cod liver oil, salt and seaweed. We discovered that where we were grazing sheep on just grass and not on a herbal ley, our sheep were deficient in minerals. As they work around our rotation and we establish herbal leys in our permanent pasture, the diversity of what they graze will broaden, and they should stop using the mineral bar as much. Animals have the natural instinct to eat what they're deficient in, to heal themselves.

The sheep's main role here is to be walking fertilisers, rather than a revenue stream in themselves. Although we do get a bit of income from culling a small amount of lambs, which helps to ensure they are cost neutral or better. What is unbelievable is the state of the wool market. We pay more for the shearer to come than we get back from the sale of the wool. Many farmers end up burning or burying the fleeces as the price is so depressed. As a result, some farmers are turning to breeds that will self-shed their wool.

Given that we need to move away from artificial fibres produced by the petrochemical industry, it just doesn't make sense. I've thought about whether we could turn it into something ourselves, but then that means creating a whole new business, which we don't have the time for, but it could be a future diversification.

Some setbacks

We've moved away from what was predominantly a two-crop rape and wheat rotation. The lack of rotational diversity is probably the main cause of widespread oilseed rape pests across the country. We've lengthened our rotation and brought in some new crops and new techniques to replace the insecticides, herbicides and fungicides we're no longer using, and we've reduced the amount of artificial nitrogen fertiliser we're putting on.

WE'RE IN THIS FOR THE LONG RUN, NOT FOR SHORT-TERM GAINS. AND WE HAVE TO ACCEPT SOME SETBACKS.

Flea beetle larvae damage in oilseed rape

Reducing insecticide application and providing suitable habitats should allow populations of natural predators to build up on the farm, to eat problem insects, such as flea beetle and stem weevil. We also plant a clover companion crop with our oilseed rape, which fixes nitrogen, provides flowers for pollinators and may divert pests from the main crop. Until such strategies are established, however, there are risks. In one of the two rape fields, we had a bad invasive weed problem which we had to treat, terminating the companion crop too. So that cost us money in seed and time. The other rape field was badly attacked by flea beetle, the larvae damage the stem and root which stunts or kills the plant. The rape crop was also affected by pollen beetle and pigeons.

It was the first time we've had it affect a crop over winter. Normally, the beetle attacks a crop when it's very young and starting to get established. It was also the most widespread infestation we'd had. Such damage reduces yields, of course, often making the difference between a profit and loss. We did manage to make a profit in 2022, but only due to the spike in the price of oilseed rape because of the war in Ukraine. In a normal year, when prices are quite a lot lower, we would have lost money.

Would the damage have been less if we'd applied insecticide? There is no way of knowing as it is a prevalent pest, even when a crop is treated, but the extent of the damage could well have been so great because we didn't. A knee-jerk reaction might be to go back to using an insecticide, but we have to hold our nerve. If we give nature a chance it will rebalance. The field also needed a rest from rape, having had it every other year in the simple rotation. In our expanded rotation, rape would be in the field for one year in every six, seven or eight, which will break the insects' life cycle in this area of land.

It will be a bumpy ride, but it's for the best. We want the natural eco-system to work with us, not against us; we want nature to provide for us, that's the end goal. We're in this for the long run, not for short-term gains. And we have to accept some setbacks. We're in it to work with nature and for nature to work with us to create better crops.

Reducing artificial nitrogen

Our goal is to reduce our artificial nitrogen use as much and as quickly as possible without decreasing our yields. In the past, at some points we'd been as high as 250 kg/ha, but generally had been closer to 220 kg/ha, which is around the average. In 2021 we reduced our application by 30% to 160 kg/ha, without our yield suffering significantly. Though 2021 was not a high-yielding year generally, we expected 2022 to be a better indicator of the effects of change. We also started trialling foliar nitrogen on four fields. Foliar applications are more efficient. If it works better, then our target is to come down below 100 kg/ha. Agriculture, as an industry, must use less artificial fertiliser. Organic and regenerative practices are leading to many farmers reducing use. The rising cost of inputs may well drive other farmers to follow suit. In 2022 we bought early and paid a relatively good price of £263/t for our nitrogen fertiliser; by the summer, prices were around £700/t – which no doubt made many think twice about their application rate.

Before we had artificial fertilisers, farmers used legumes (for example: clovers, beans and peas) to take nitrogen from the atmosphere and fix it in the soil. We are using these leguminous companion and break crops, and we are working with our agronomist, Graham Chester (Assured Agronomy), and regen consultant Ben Taylor-Davies, to help us reduce our usage of applied nitrogen.

We told Graham we wanted to adopt regenerative practices. At that time, he didn't have many clients who were regen. Agronomists, like so many farmers, are products of the conventional education and training

Why artificial nitrogen is so damaging

The historic reliance on nitrogen and its overapplication is severely damaging the environment. The unused nitrogen runs off and pollutes watercourses, causing algae blooms. Microbial action in the soil converts it into nitrous oxide, which is a much more potent greenhouse gas than carbon dioxide, and elevated nitrate levels in drinking water are linked to increased risk and incidence of diseases, including cancers.[3]

system, and they also need to relearn. We said, 'We want to go in this direction, will you come with us.' And he said, 'Yes.' Sometimes we'll push him to go further, and other times he's suggested we can do something better or in another way – it's a partnership.

We use Ben Taylor-Davies for regenerative prescriptions, especially for nutrition and nitrogen reduction. His recommended mix in 2022 was 130 l/ha containing Brineflow ATS nitrogen, 20 l/ha supplemented with TERRA FED 2.5 l/ha, HUMA GRO at 0.5 l/ha, and 3 kg/ha of EPSOTop – tailored to apply with the nitrogen product to make it much more efficient and reduce our overall nitrogen application rate. TERRA FED is a carbon-molasses product (it is applied to maintain the carbon : nitrogen ratio in the soil, too much nitrogen compared to carbon means take-up is reduced). HUMA GRO is based on humic acid (it is applied to help the nitrogen attach itself to the plant, but also to the soil, reducing leaching of excesses into watercourses). EPSOTop provides magnesium and sulphur. We're also trialling a fish product which acts as a sort of multi-mineral supplement. Carl hates it as it smells and has a tendency to block his sprayer!

We conduct plant-sap tests on the leaves to measure any trace elements that might be missing in a crop (for example: boron, magnesium, manganese). We analyse the tests and apply products to supply what's deficient. It is important to test; if not, you might overapply one mineral and lock another one up. It's a very delicate balance.

Shimpling Park Farm

During the autumn of 2021, Dad and I travelled down to Suffolk to visit John Pawsey at Shimpling Park Farm.

ANNA: John, please introduce us to your farm.

JOHN: We are Shimpling Park Farm, which is about 8 miles (13 km) south of Bury St Edmunds in Suffolk. We are a mixed farm. We've got sheep, and the rest of the rotation is cropped with cash crops. We're all organic. We converted in 1999 so have been organic for over 20 years.

ANNA: Why did you become organic?

JOHN: We were conventional at the time. I was doing all the spraying on the farm, and I was increasingly mixing up and applying all the chemicals, and it just felt completely out of control. I know everyone has their hare story, but I had this moment where this hare leaped up in front of my sprayer and ran underneath the booms and got covered in the spray. And it just sat on the edge of the field licking itself. This was something macro I could see, and I suddenly realised I was spraying everything else as well, not just the soil but the drift going into the hedgerows, the field margins and all the micro things that I couldn't see. I just realised that the easiest way for me to stop feeling bad about it was to stop doing it.

John Pawsey (left) with Anna and Andrew Jackson.

ANNA: We have three in our operation. How many people do you have on your farm?

JOHN: We have three people on the arable team, that's not including myself, and we have a shepherd. We're farming in the region of about 4,000 acres (1620 ha) with a relatively small team. For specialist jobs, we have contractors come in to help us. And then we have lots of people who rent buildings from us for doing lots of different things on the farm. I've got a wages book from my grandfather's day listing 54 people. But if you look at the amount of people who are actually working on the farm, including in the rented buildings, we've still got about 54 people. It still feels like this sort of huge community, which is really important to us.

ANNA: Do you think being regen and organic has introduced community, or did you always have that?

JOHN: As far as I'm concerned, it's all about relationships. It is not necessarily just an organic thing, but before that, it was almost like we weren't producing food. We were just producing this commodity. We went out and did the high-input, high-output type farming system, chucked it all in the grain store, sold it to a merchant, and a lorry would come and pick it up. I would have absolutely no idea where it was going. But when we went organic, we sort of went niche, so we had to start building relationships with people. Relationships are incredibly important because very often the last thing you talk about is price. The first thing you start talking about is what everybody needs from the relationship, which is the most rewarding thing. And if you can agree on a price that is fair to everyone, then hey ho, that's great news.

ANNA: What's your family's role in the farm life?

JOHN: The family's role is really important to me. My wife, Alice, and I, first of all, we can't work together, so we have to have separate jobs, and then we work very well alongside each other. We work very

I JUST REALISED THAT THE EASIEST WAY FOR ME TO STOP FEELING BAD ABOUT IT WAS TO STOP DOING IT.

differently. I would be really keen for any of my children to come onto the farm in the future. The most important thing for me is just having the right people on the farm, that they're people I want to be working with. We put a huge amount of time into building relationships with the people who work with us on the farm, not just the full-time people but the contractors as well. It's about getting the best out of people, which you do when they feel that they're actually part of the whole thing and are contributing. You show them love and gratitude and explain what you're doing and communicate. So, it's not all about immediate family – it's about all the different people who come and work with us.

ANNA: What food are you growing here?

JOHN: We're producing lamb and we're growing a lot of combinable crops, so that is crops that go through our combine harvester: wheat, oats and barley; we're also growing beans. We do a lot of bi-cropping so growing two crops together to mitigate the risk of having too many weeds in the crops or pest attacks or fungal diseases. We grow very diverse leys that feed our sheep and also build fertility in our soil. We grow some novel crops like lentils and chia for Hodmedod. This year we've added quinoa. Sometimes we're only doing small areas, but it's one of those things that, as a farmer, keeps you interested. Agronomically, it is really exciting.

ANNA: Say I am a conventional farmer, I only grow rape, wheat and sugar beet. Why should I be growing more crops?

JOHN: You probably shouldn't be if you're happy with growing three crops. I'm just not happy with growing that. First of all, it's about learning new stuff and learning how to grow different things. You can be the best sugar beet grower in the country and get huge recognition for that, and I applaud that of course, but it's just not something that I particularly want to do.

ANNA: How does having more than three crops benefit the land?

JOHN: Actually, it's a really interesting question because, thinking about climate change, if you have lots of different crops in your rotation, you're mitigating to a certain extent against weather events because they're all going to react differently. It's about spreading your risk by adding diversity, which brings resilience, both for you and for the nature on your farm.

ANNA: Who are your customers?

JOHN: We sell our oats through a company called Organic Arable, which is a farmer-owned business that I'm involved with. The spelt goes to Sharpham Park, the quinoa to the British Quinoa Company, chia to Fairking and, along with lentils, also to Hodmedod, lots of different people.

ANNA: And for the organic lamb?

JOHN: That's down to my lovely wife, Alice, because I'm a useless salesman. She has great relationships with the local butchers and individual customers who buy direct.

ANNA: So everything goes into the human food chain?

JOHN: Before we went organic, we were growing pretty much all our crops for the animal-feed industry: barley, wheat, beans, sugar beet. Now pretty much everything, depending on the quality we get at harvest time, goes into the human food chain. One of the most amazing things I have to say about our food now is that it's feeding people. Because of the way some of the companies work, I can go into a supermarket and see a packet of something I've grown. For instance, Hodmedod have been brilliant in this way; I can pick up a packet and see 'Chia grown by Alice and John Pawsey.' I mean, it's just a wonderful, wonderful feeling to have that sort of direct connection, with our names on the packet – something I think most farmers do not get to experience.

British agriculture, with organic, with regen, with high welfare, has got fantastic and different stories to tell. And as farmers, we're very bad at telling that story. In years gone by, there would be names of the farms and farmers on packets in our local shop; and I see a future where food provenance and traceability will give people a connection with food and the farmers who produced it.

I SEE A FUTURE WHERE FOOD PROVENANCE AND TRACEABILITY WILL GIVE PEOPLE A CONNECTION WITH FOOD AND THE FARMERS WHO PRODUCED IT.

> ## Hodmedod – making connections
> Hodmedod was founded in 2012 following the successful Great British Beans trial project to stimulate demand for pulses. Nick Saltmarsh, Josiah Meldrum and William Hudson were trying to reintroduce people to fava beans, grown here for more than 2000 years but now largely forgotten and struggled to find a wholesaler from which to buy. So, they set up their own – Hodmedod, which now connects over 35 farmers producing a range of pulses, grains and seeds and much more with customers directly and via trade stockist. They feature their farmers stories on their website and on packaging. (See also Josiah's contribution to Chapter 8.)

DAD: Are we are we in danger of overcomplicating the message when we talk about organic, regen and things like the red-tractor label? Are there too many stories for the customer to understand?

JOHN: It is a complicated place out there as far as messaging to customers is concerned, but I still think it's not an excuse for not doing it. You've got to give it a go, and there will be some people who do well in that space and some people who don't. But I just don't think it's a reason to say that you're not going to do it. We've benefitted enormously from the connections we've made.

ANNA: What does organic mean to you? And what are the differences with regen?

JOHN: Knowing my farm is not sprayed with chemicals actually means a huge amount to me. I can't really put my finger on why that is, other than I spent a huge amount of time sitting in a sprayer feeling uncomfortable about it. I just prefer my farm to be growing without chemicals.

The difference between, in my view, organic and regen is that when you switch over to organic, there is a set of rules you just have to follow. You have to turn off the old system and turn on the new system, and it immediately makes you become inventive; it makes you more creative, and I found that really engaging. It does mean that the risks are higher at the beginning of your conversion, especially if you haven't got good advice. But I found it liberating. I knew what I couldn't do, and I had to immediately find solutions to those problems. I really enjoyed the transition.

ANNA: What were the problems?

JOHN: Oh gosh, we haven't solved nearly half of our problems, but that's the exciting thing about it. You have to be inventive and it never stops. Every year is different. I would say that after 20 years, I'm beginning to relax more, and I feel that our soils are improving. I've never felt more excited about how things are going, as far as the farming part of the business is concerned. Being on that learning curve is still a hugely enjoyable ride.

The worst problems happened when I didn't keep my eye on the organic ball. We started off in a relatively clean farm, having sprayed it with herbicides for scores of years. I didn't really understand what the weed problems were going to be like at that point; for example, I let some wild oats go, and it took us 7 or 8 years to bring those fields back round again.

ANNA: How did you bring them back around?

JOHN: Through different agronomic methods; for instance, with wild oats you should leave your fields as stubble for 3 or 4 months after harvesting because the wild oat seeds are eaten by birds and mice. If I'd just pulled that one wild oat in that first year of being organic and kept on pulling them as they came up, I would have been fine.

ANNA: Going back to marketing, do you think you have an upper hand being organic?

JOHN: It's an interesting question because organic is sort of quite divisive. Some people think it's expensive, so that can put them off, but I personally don't think it is expensive in the long term because of all the benefits that organic delivers. We possibly err on the side of

ORGANIC CONVERSION IMMEDIATELY MAKES YOU BECOME INVENTIVE; IT MAKES YOU MORE CREATIVE, AND I FOUND THAT REALLY ENGAGING.

Not sowing my oats

When I first came back home to farm, my grandfather was very ill, and I used to cook his lunch every day. I used to make a casserole in the morning and serve it up at lunchtime. I came in for lunch one day and he said,

'Did you see those wild oats in the field before you come down the drive?'

And I said, 'Yeah.'

And he said, 'So what are you doing in for lunch then?'

And he meant, 'Well, you should be out there pulling them out.'

marketing Shimpling Park Farm as a destination and then the fact it's organic – rather than we're organic and then we're Shimpling Park Farm, but people are always interested in the human story, which is why I think that the regen movement has a huge story to tell, and I'm not sure that in marketing terms it's successfully telling it at the moment.

ANNA: This confuses me a little: you are organic but you also have regen practices. How does that work?

JOHN: I think you've got it the wrong way around because organic has to be regenerative to work. If you're not using any artificial fertiliser or any chemicals and you're still producing profitable crops and your soil is getting healthier, it has to be regenerative. I think there's some kind of suggestion that organic is not regenerative, but it's massively regenerative. It just has to be. What's really interesting about the regenerative movement is that it's using all the words and all the terms about soil health that organic farmers have been using for scores of years. I almost feel that it's been hijacked to a certain extent. I would say where we're regen first and organic second.

ANNA: How many sheep do you have and how do they work on the farm?

JOHN: We have 1,000 New Zealand Romneys. We've got a small amount of permanent pasture. The rest of the time they go around these fertility-building areas on the farm, leys that build fertility in the soil, and they graze them for us. Before we had sheep, we used to go with a tractor and a mower and just mow these leys and then leave them as a mulch on the soil, costing a huge amount of money in time and diesel. We decided to bring in sheep to do the work for us. Basically, I see them as little composting machines going around the farm, just doing on-farm composting.

ANNA: And this makes the sheep happier, right? Because they've got a variety of food in their diet. You know that they're not seeing the same place all the time. They're having a bit of an adventure.

JOHN: Well I love the way you put that. From now on, I'm going to think of it like that. It's like going, 'Oh my God, this is a new field. That's fantastic. Have you seen that lovely bit of chicory down there?' I hope they do. I mean, you know, rather than being restricted to a bit of permanent pasture, they are going around the farm, seeing the countryside.

ANNA: How different is your system to an intensive system?

JOHN: Well, they're not inside. They stay outside 365 days of the year. They are grazing herbal leys. I just think it's as near as you can get to an extensive, holistic sort of grazing system. Some systems have them come inside to lamb. Ours lamb outside. We've selected a breed that can really look after itself and look after its lambs naturally, which is really important to us. It's completely different, to be honest, I don't think sheep can be farmed intensively.

ANNA: Can you measure the effect the sheep have?

JOHN: We have the sheep on our farm, but we also farm for other people and we don't have sheep on their land, so I can compare how the farms without sheep are doing with my own farm with them – we've only had sheep on the farm since 2014, and actually both farms are yielding pretty similarly. It will be interesting to see what animals bring to the rotation. I don't think it has been long enough yet, so I can't really say whether or not the animals add a huge amount to our rotation.

I WOULD SAY WHERE WE'RE REGEN FIRST AND ORGANIC SECOND.

My gut tells me that they are, but at the moment there's not a huge amount of difference between us and the farms we're farming without sheep. Having sheep has enabled us to extend our rotation, which I think is important, and they enable us to deal with certain weeds on the farm, which is also important. But I think it's a case of 'watch this space'.

ANNA: And they bring additional income?

JOHN: Correct. It's another bit of diversified income for the farm. Last year, financially, they did incredibly well. This year it's been very dry, so they've not done so well. But it's swings and roundabouts with all the different things that we do on the farm, and to have another income stream brings a little bit of risk management to an enterprise.

ANNA: On our farm we've worked out the target number of sheep that we want to stock. A number that, we hope, is both manageable and sufficient to graze the amount of land that we have. How do you know what's too many or too few?

JOHN: It's quite difficult to judge how many sheep you need. Roughly, we have about eight ewes per hectare and that suits our system very well. All farms are different, but we've got quite a heavy soil which grows a huge amount of grass and clover; by the end of some seasons, we have more than the sheep can eat. I think you've just got to look at your local situation, understand your land and get advice from others keeping sheep locally.

ANNA: What's wrong with our current food system and how can we fix it?

JOHN: It's a big question, isn't it? The problem that most of us farmers have is we have little connection to the market. I believe that there are many more routes to market than most farmers realise. We've got to engage with our customers more. There's no doubt about that and doing so will create new opportunities.

ANNA: What is the future of farming? What do you envision or what would you like to happen?

JOHN: For me it's more and more about diversification. It's about looking at all our assets and working out what we want to be doing with them and how they can bring new income streams into the farm. I don't think it's all about food production. More generally, I think it's about creating new opportunities for farmers and for people working within farming businesses, and that's incredibly exciting. The loss of the Basic Payment Scheme, I think, will create huge opportunities. It will make us leaner, it'll make us more inventive and, ultimately, it'll be much better for our industry. There are really, really exciting times ahead. And we haven't even started to talk about robots!

ANNA: There seems to be an expectation out there that technology, be that robot farmers, precision agriculture, vertical farming or synthetic meats, is going to solve our problems. What do you think about the ever-increasing force of technology?

JOHN: We had a conversation about robotics with the guys on the farm the other day, and they said, 'Oh, my goodness, you know, we're all going to be replaced by robots.' My feeling is that, yes, probably robots will come onto the farm, but it'll allow us to be better practitioners, to have a better and different focus on the agronomy that we haven't been able to do because we've just been sitting on tractors and looking at this stuff from on high in a cab.

We're going to be using technology more to solve some of our problems, but we're going to be able to have a much more intimate relationship with our soils and plants and all those agronomic aspects that we've probably not been engaging with well enough.

DAD: One thing that just about beat me when I was having my spell in organic production was the dock plants. Once they go to seed, you've got a massive problem. How do you control them?

JOHN: One of the difficulties we've had with not having any kind of herbicides is weeds like creeping thistle and dock. Both hate competition. If you manage to get a crop well-established early, generally in the autumn, you can keep on top of them. We've managed to keep the numbers down by just growing more competitive crops like spelt and winter oats. Correct cultivation helps with them as well – cultivation at the right time to leave those dock roots drying out on the soil and to get underneath the rhizomes of those creeping thistles.

There's a great farm manager at the Sandringham Estate, Keith Banham, who did quite a lot of work with low-disturbance subsoiling. Getting the subsoiler wings just underneath those rhizomes. The control of creeping thistle was pretty phenomenal.

We've got a thing called a CombCut machine which cuts them off before they flower or set seed, which also helps weaken the plant. I think what's really interesting when you're farming without herbicides is there's no one solution. You've got to be doing a little bit of everything to keep on top of these things.

I've come to the conclusion that the appropriate cultivation for that particular crop for that particular season, if really well considered, is less damaging to soil fungi but also earthworms. This year has been incredibly dry, so we are doing some ploughing, but we are ploughing no more than 4 inches (100 mm) of our 12 inches (300 mm) of topsoil. All we're doing is inverting it to deal with weeds. But only because it's the kind of year that you can do it in – it's so dry that we're still leaving the majority of our topsoil completely undisturbed and obviously the subsoil too.

I think farming without herbicides and doing no cultivation is going to be very difficult unless, with robots, something comes along to target pernicious weeds. What we're focused on is getting the timing of those cultivations right and making sure we have the right kind of cultivation to deal with the problems.

ANNA: I think the other thing is the right depth; we've been cultivating for too many years, too deep with tractors that are just too powerful.

JOHN: That's true. I've got these two young guys on the farm who've never ploughed before. So ploughing at no more than 4 inches is completely normal to them. Whereas some of the older people on the farm find it very difficult not to plough as deep as they possibly can because that's what has always been done. If you think about it, a horse would never have been able to pull a plough so deep.

DAD: We asked a gentleman who worked on our farm for 60 years how deep the horses ploughed in his day. And the answer was: 4 inches. My question is would you be better off making multiple passes with a cultivator?

JOHN: This year it's been so dry where we've broken leys without using the plough, that we've had to go with four cultivation passes to try to kill it, whereas ploughing at 4 inches, we've inverted the soil and we've followed it with just one cultivator pass.

DAD: Saving time and diesel with a shallow plough.

JOHN: Some fields here probably haven't been ploughed for 6 or 7 years because the weeds in the field or the crops we've grown haven't required something as drastic.

ANNA: A lot of people demonise the plough in the regen world. We don't use one anymore. We sold our plough. But actually, if we want to get rid of chemicals, a plough might be needed in our system. How do you do that in the least destructive way possible for your soil?

JOHN: It's all about conditions and, as you said, it's all about depth. Ploughing very shallowly and leaving most of your topsoil undisturbed and certainly your subsoil undisturbed, to my mind, is a fair trade off for that fertility-building period. If you are ploughing in very wet times when your worms are near the surface, you're going to chop through a lot of worms, but if you plough when it's dry and the worms are deeply buried in their burrows, then it's a totally different scenario. Considering the fungi in the soils and the mycorrhizae, if you're only ploughing the top 4 inches, and you've got 12 inches of topsoil, I'd argue that is a fair trade off as well.

It's all about understanding your soils, understanding how you're building fertility and understanding the balance of your soil health. So long as you're monitoring it and soil health is going up, as far as I'm concerned, it's okay.

ANNA: Are you ploughing every year?

JOHN: No, we rotationally plough. We will plough once in the rotation, but sometimes we don't have the need to. To summarise it, every field has a managed approach as far as cultivation is concerned. There's no blanket approach of ploughing everything or not ploughing everything. We go into a field, and we do what is appropriate for that field at that time.

Being a young female farmer

I get asked quite a lot whether it is difficult being a farmer who's a woman. I am certainly in the minority. I went to a conference the other day and 300 people were there, and I was one of two female farmers. But you know what, if I let it bother me, then it will bother me. Whereas if I'm successful in what I'm doing and focus on being me, hopefully I will be a role model for other future female farmers.

I think more women need to get into agriculture. All the women I've met in agriculture have been doing incredible things. I feel it's happening but just too slowly. There does seem to be more women in regen than in conventional farming. The gender splits at the Cereals and Groundswell (see page 275) events seem to be testament to that. I've heard it suggested that women have a greater respect for and ease of connection with nature. That they are drawn to the regenerative. I can't speak for other females, but I love being in nature, the variety regen brings and the knowledge that I'm being proactive.

Physically it's been harder than I expected. I've always exercised, but I've found my personal strength is often tested. However, I just think or find ways around it; if I can't open a valve, I use a length of metal pipe and get some leverage. I'll build up my strength over time.

What concerns me more is how the young and new entrants can get into farming, be they female or male. It's just so hard because there aren't many farms available to rent or to buy. I think it's ridiculous. To land a tenancy, you have to bid, and how does a new entrant estimate what to offer. Then they have to buy the equipment and set everything up. I'm really lucky that Dad has a farm.

It's very difficult. I think there definitely needs to be more support from the government for new entrants. Look at Adie (Chapter 5) and Ben (Chapter 6), they're doing incredible things with just their own resources. I'm lucky that I've got financial back up. Adie's making her wage. She's living day by day, figuring it out. That is scary and it's what new entrant farming is like.

I certainly couldn't go back. I think this is the most content I've ever been in my life. I've got my dog, Luna; she is incredible. I've got my sheep; they're incredible. I'm working with my dad and we're getting on really well.

Luna, helping with harvest.

Harvest 2022

The second field of oilseed rape, the one damaged by flea beetle over winter, did not do as well as expected. For rape, we'd want to be near an average of 4.5 t/ha, but we were hitting 2–3 t/ha in the best areas of the field and averaged 1.7 t/ha. We struggle with rape, to get it in early enough, because there are only three of us. It's tricky to move someone onto a drill when we're still harvesting wheat.

Along with most other farmers in the country, wheat is our bread-winner. Wheat is the crop that reliably makes money and keeps us going as farmers. It makes up for the other crops and allows us to experiment with the rotation. It would be really cool if some of the other crops performed as powerfully as wheat. The obvious question might be, why not just grow wheat and make more money? Well, we don't want to grow a monoculture of wheat because that's not good for nature or wildlife, and the yields would go down over time, as the soil became exhausted and diseases took hold. Our break crops don't make us a lot of money, but the wheat certainly wouldn't be as good if we didn't have our break crops in the rotation. It's all about balance.

The wheat performed strongly. Across the fields we were hitting 11–12 t/ha and spiked at 13.68 t/ha, averaging at over 10 t/ha. We're getting these results while reducing our artificial nitrogen year on year and being regen now for 8 years. One of the main things holding others back is fear of a drop in yield. Which is understandable as the whole system is set up to maximise production and to see a good farmer as one who pulls

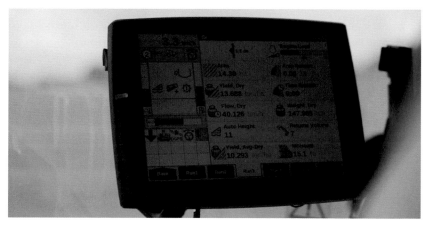

Top yield 13.68 t/ha, averaging at over 10 t/ha.

in massive yields. We can still get yields of 11–12 t/ha farming regeneratively, it's doable. So there is no excuse really.

When harvesting, we maintain our controlled traffic farming, which means that when we drive any vehicles on the field, we stick to the tramlines, so we minimise compaction of the land. If we're unloading, we only do so on tramlines or off the field.

In lots of harvesting videos, you'll see the tractor and trailer drive alongside the combine and they unload into it as they go – making it look exciting! If you have a really big farming operation, you might have to do that because of time. But we don't need to; we want our fields to be full of living roots and quite squishy. A similar feel to walking across

ONE OF THE MAIN THINGS HOLDING OTHERS BACK IS FEAR OF A DROP IN YIELD. WE CAN STILL GET YIELDS OF 11–12 T/HA FARMING REGENERATIVELY, IT'S DOABLE. SO THERE IS NO EXCUSE REALLY.

a grass field, the land giving underfoot. A field that's been ploughed for years, especially when traffic movement isn't disciplined, can get hard and compacted, and then roots struggle to grow down into the soil.

Another visible indication of our regenerative practices is stubble that's left high. It provides a wildlife habitat, protects the soil until a cover crop grows through, and leaves roots in the ground. We'll then drill straight into this with the next crop in the rotation.

In an exciting development, we took some of the wheat just harvested and straightaway direct drilled it into one of the rape fields containing the remaining clover companion crop. The plan was to get the wheat established early enough for us to bring the sheep in over winter to graze it and the companion crop off – to promote vigorous wheat growth in the spring and of course to bring all the benefits of the sheep to that field. This is an established practice in warmer areas of the UK but one we've not been able to do yet with the cooler climate further north. By starting the wheat early, we hope to give ourselves the best chance of trying this out.

For 2023 we also brought in soft wheat for the first time, which is typically used to make cakes and biscuits. We blended 10 varieties together. The diversity of varieties, in theory, will give greater resistance to disease, making it harder for anything to spread from plant to plant than with a monoculture approach.

We use Humber Grain to market our wheat. Once it's sold, the wheat is loaded into a big lorry, and that's the end of its journey for us. We watch the wheat price fluctuating and try to sell at what we think is a good price. It's a bit like gambling. Ideally we'd like more connection with customers. I'd really like to sell wheat locally, perhaps mill on site or in the vicinity. But for arable crops it doesn't really work like that, not like meat and vegetables sold in box schemes and at the farm gate. Perhaps if we could connect better, then we might be able to achieve a higher price in recognition of our regenerative practice.

Currently, even though we've farmed our land as chemical free as we possibly can and are on our regenerative journey, we wouldn't get a premium on this. There's no premium for regenerative wheat or regenerative products in general. There are a few companies that are starting to appreciate where it's coming from and how it's grown. In the US, I understand, they are trying to test the nutrients of the wheat to evidence its higher value. Currently there is no mechanism in the UK through which we could receive any more money for our regenerative wheat. It just gets mixed in with all the other wheat crop. It's unfair in a way that we've paid quite a lot of money to make our systems better for the environment.

At the moment we have conventional farming, regenerative farming and organic farming, and we're in this middle ground between conventional and organic, and nobody really knows what to do with it, in terms of business. Conventional will be sold at certain prices, organic will be sold at premium prices, partly because of lower yields because they're not using chemicals, and we're in the middle, trying to get the same yields as conventional but trying to use less chemicals.

We are choosing to be regenerative but with no premium like organic. We're choosing to do that because Dad and I couldn't sleep at night with a clear conscience, knowing that we haven't done everything we could. This is our patch of land. We want to do everything physically possible in our control to be as environmentally friendly as possible. If that means we don't get extra money for producing better food, then it doesn't matter because we still want to do it anyway. We'll do it without extra payment. It's for everyone, farming. Everyone eats, everyone lives on this planet.

Running the numbers

As I'm sure most farm businesses do, we start the year with a budget and a forecast of how much each of our crops may yield. After harvest and when sold, we reconcile the yields and net margins for the year as part of our financial management.

Wheat performed as expected, or better because the Russian invasion of Ukraine drove the wheat price up. Grass seed also came in fairly well. Oilseed rape made a profit but as usual not as much as we'd like. The vining peas should have done better, but because they are grown on rented land which ate into the margin, they only just made it into the black.

Then it comes to loss-making crops. We tried meadowfoam; we were some of the first growers in the country to do so. It's a very shallow-rooted crop, and it was a very, very dry summer, and we lost money on it. As new varieties are developed, we may grow it again, but not for now. It was a nice crop, but difficult to harvest.

Then quinoa, it ticks all the boxes on paper. It's a break crop. It's spring sown, so you take a cover crop over the winter months, graze it off and then establish the quinoa. We've grown it for 3 years. You can't say we haven't tried, but we've lost money for the third year on the trot, so we probably won't be growing it again. Whoever grows it in the future with

| Wheat +£950 | Grass seed +£650 | Oilseed rape +£410 | Peas +£2 | Meadowfoam −£300 | Quinoa −£610 |

The 2022 harvest (left to right): wheat, grass seed, oilseed rape, vining peas, meadowfoam and quinoa (net margin £/ha).

our increasingly hot summers will need more money than we were paid per tonne to make it work.

As a whole in 2022, we only made a profit as a farm because of the BPS that we receive from the government. As we're not on rented land, we can afford to be experimental, we have no landlord to satisfy. We are basically trying to push regen as far as it will go, but with experimentation comes risks and losses, and this year we made a loss on our cropping.

Since going regen 8 years ago, profits have roughly stayed the same as when we grew conventionally. If we'd stuck with a steady rotation all the way through, profits would have risen. Our experimentation has brought some failures, and that has dragged average profits down. However, taking wheat in isolation, the yields haven't dropped since going regen, and we've seen the margin go up significantly because of the huge savings we are making on establishment costs, with greatly reduced input levels. We are making more money growing wheat now than when we farmed conventionally.

WE ARE MAKING MORE MONEY GROWING WHEAT NOW THAN WHEN WE FARMED CONVENTIONALLY.

Our wheat hasn't had a yield dip since we went regen because we've done it slowly. We haven't taken huge risks with the wheat. That's definitely something to hearten others, particularly those on rented land. It is possible to change and not to take a hit on profits. If we cut out the loss-making crops, we'd be in clover.

There is a 7-year rotation, spread out across our whole farm, with each field having a different crop in it. We've changed this for 2023: dropping meadowfoam and quinoa and adding oats instead. We're still looking for the elusive, money-making break crop and hope to add another so we will be back on a 7-year rotation.

Reflections

Dad's nearing retirement age and could be slowing down, but regenerative farming has given him a new lease of life. He's not always been passionate. He's gone from, 'I don't know if I want to carry on' to 'I'm loving this. I'm doing this every day.'

He'd been farming the same way his whole life. Since he discovered regen farming – he literally just read a book and was transfixed – it's like he's a teenager again. He's on YouTube till late at night watching regenerative agriculture videos, he's reading books, he's a member of BASE-UK, and he's meeting all these other farmers who are also going regen.

The regenerative approach is so much better for mental health in the industry – less tractor time, more walking, more smelling, more seeing what's going on and more communication with other humans, all of which is so important. The smell and touch of the soil is known to be beneficial and to lift mood.[4]

As an industry, we need to move away from chasing big yields. That's what farmers were expected to do. Even Dad gets excited by high yields. I think that's because it's ingrained – supposedly the higher the yield you have, the more money you make. I don't think that's what our goal should be. We should be looking to grow higher quality over quantity. We need to scale back to what our land can sustainably produce. We can have higher profits off lower yields with lower inputs and better food.

In most cases, farmers are price takers and not price makers. Dad recalls that only once in his life was he able to set a price for his crop, when he had a good yield of organic parsnips and few others did. We do have a Birdseye growers negotiation committee for our pea contract, and we like to think that we get a fair settlement, although we never think it's as fair as it should be. For everything else, we are price takers because the commodities are sold on the world market. The quinoa was grown to a contract, and obviously we thought we could make money. But then weather variability kicks in with lower rainfall and high temperatures, the yield is lower, and all of a sudden, the price wasn't enough to make a profit.

The climate is one of the main reasons why we've gone regenerative, because there is so much extreme weather. We need to make ourselves more resilient to weather shocks in the future. Healthier soils absorb more water and are more drought resistant.

It all comes back to the soil. Within Europe as a whole, we haven't historically had the weather extremes that the Americas and Australia have had. We have lost topsoil, but not on the scale that they have in the American Dust Bowl.

What is more apparent is that we've lost organic matter. Our sandy land, which naturally is low in organic matter, is now very low. And some of our better-bodied land, which should be good, is not what it used to be. Organic matter readings from the hedge bottom and the field should be the same. But the fields are much lower in organic matter because they've been cultivated for the last century or two. It is the lost organic matter that holds the water, holds the nutrients and holds the fungi. Since we moved to direct drilling, we've stopped disturbing the soil and seen a slow build-up of organic matter again. We've raised it about 1.5% in 10 years, which doesn't sound a lot, but it's going in the right direction. All farmers should be measuring their levels of organic matter. Are they going up or down? If they're going down, they need to think about what to do about it. What we're trying to do is prove that it's financially viable to build our organic matter levels and remain profitable, even to be more profitable.

Farmers are having to relearn their trade. Dad and I are in there with them and we're learning as quickly as we can. It's a really good community.

WE SHOULD BE LOOKING TO GROW HIGHER QUALITY OVER QUANTITY. WE CAN HAVE HIGHER PROFITS OFF LOWER YIELDS WITH LOWER INPUTS AND BETTER FOOD.

Regenerative farming is not just about the environment, it's about working together, working with your neighbours and learning off each other and from farmer-led groups like BASE-UK.

Connections, as John Pawsey noted, are so important – reaching out beyond the farm to form mutually beneficial relationships, not just to share ideas but to bring in new people and resources. Since going regenerative, we have tried to create partnerships. We do this with sheep; we have extra sheep come on some of the fields when our flock isn't sufficient to graze everything off. A local shepherd comes in and grazes sheep for free. We get the fertilisation and he gets fatter lambs. We've also done a similar thing with bees; when we had the meadowfoam, we messaged a beekeeper, offering him access to a field full of flowers needing pollination. We're increasing our yield because the bees are pollinating, and his bees have access to the nectar, resulting in more honey for him.

A lot of what we as a regenerative farming community are doing lacks research support. Much of the science is new, and there just hasn't been the chance for the universities and research centres to catch up. Conducting trials and gathering sufficient data to robustly prove results takes time of course, but there is also a question about funding and will. Research in the agricultural industry over the last 50–100 years has often been funded by big business, the agrochemical industry. We live in a capitalist society

and, maybe like the US, there's more money to be made by the government from the profits of the fertiliser and chemical companies than from the farmers.

The move in the US and now over here is for farmer-led research from the bottom up, supported by forward-thinking research centres. No industry is going to fund research that proves we don't need their inputs. Farmers are doing the research themselves, and they're actually finding that it can be done. The Americans can do it, the Argentineans and Brazilians can do it, which gives us the confidence to have a go.

People are disconnected from their food. If you went back 100 years, most people would either work in the countryside on a farm or have a relation who did so. So there was much more of a connection, but now farmers only make up about 1% of the population. There's a bit of coverage on television with *Countryfile* and more recently *Clarkson's Farm*, and of course *The Archers* and *Farming Today* on the radio, but farming is far from being in the popular consciousness, and regenerative agriculture generally features very little. We are trying to reach out; we have school groups come to the farm, and I do Farmer Time where I chat to a group of kids in a classroom once every 2–3 weeks on a video call. I basically introduce them to what we are doing that week. It's why I do my Tik-Tok videos and share stuff on social media. It's not that I want to be an influencer because I really, really don't! I just want to share what we're doing as farmers and to communicate with the general public about things like how their food is made, where it comes from, what's happening with farmers and why there's a shortage on the shelves – keeping this open conversation, which I think is really important.

PEOPLE ARE DISCONNECTED FROM THEIR FOOD. IF YOU WENT BACK 100 YEARS, MOST PEOPLE WOULD EITHER WORK IN THE COUNTRYSIDE ON A FARM OR HAVE A RELATION WHO DID SO.

Dad will never retire. No, no, no. He will not retire, which is fine by me. I don't want him to retire. He can farm with me as long as he wants. I'm more than happy! I think eventually the natural generational progression will mean that I will take over more responsibility on the farm. Hopefully by that point, I'll have created some diversifications, so that it can pay me a full-time salary – that's one of the end goals. The 10-year plan is to still be farming and have lots more sheep.

During filming we were both asked what our impossible dreams were.

DAD: To grow 10 t/ha of wheat using direct drilling, not disturbing the soil, no phosphates, no potash applied to the soil, no insecticides, no fungicides and less than 100 kg N/ha.

ANNA: To use no chemicals whatsoever, including glyphosate, and to have a closed cycle on the farm, with no inputs coming in and only our produce leaving.

Actually, they don't seem so impossible; every day we're moving a step closer.

Interlude III

Subsidies, Brexit and trade

VICKI HIRD

Farmers and growers in many regions and countries receive public financial support or subsidies. Support can be purely to keep food supplies flowing, maintain rural economies and employment, or deliver public goods like nature protection or recreational access. Of course, that support, or changes to public support, may also serve political purposes. They may also have unintended consequences.

The Agriculture Act 1947 and Common Agricultural Policy

Modern UK subsidies were established by the postwar Agriculture Act 1947. The act established guaranteed prices, set annually by what was then the Ministry for Agriculture, Fisheries and Food (MAFF) and the National Farmers' Union (NFU). There were specific subsidies and capital grants to support a range of government policies, such as the ploughing up of non-agricultural land, increased fertiliser use and the extension of hill farming. Payments were increasingly linked to production.

In 1973 the UK joined the European Union and implemented their Common Agricultural Policy (CAP). The CAP is a set of laws created in 1962 to establish a common, unified policy on agriculture in Europe, under which farm subsidies are paid across a huge, diverse region. The CAP is a beast of a policy, and the budget at the time the UK left the EU amounted to £3.2 billion of support in the UK.

The CAP evolved over time. Moving from direct subsidies linked to produce, which led to the infamous butter mountains of the 1970s–2000s, when production far outstripped demand, to the Single Farm Payment introduced in 2003, which subsidised farmers on a per-hectare basis. The addition in 2014 of rural development payments brought some limited incentives for greener farming. The area-based Single Farm Payment, by design, gave 80% of support to 20% of farms. The bigger the farm, the more money was received. There was a huge loss of smaller farms and major amalgamation across the EU, but it is also true that far more farm holdings would have been lost without the CAP, so in that sense, it served a rural social function.

The CAP and its predecessor subsidies in the UK had many faults, not least in overseeing a huge shift to agricultural intensification, with the associated harm to nature, ecosystems and livestock welfare.

By the time of Brexit, around 80% of the UK subsidy was provided as direct payments under the UK's Basic Payment Scheme (BPS), based broadly on how much land was farmed. A further tranche of the CAP budget was spent on rural and environmental programmes such as England's Countryside Stewardship (CS) scheme. It ensured many thousands of small and family-run farms across the EU and in the UK stayed in business, but overall, the evidence showed the CAP was not working well and certainly was not supporting nature-friendly farming.

Post-Brexit farm support

One significant, possibly positive but as yet largely unmeasurable, impact of Brexit was the opportunity to decide for ourselves how we, as four nations, support our farmers to make better use of the land and to regulate farm-

ing. The Agriculture Act 2020, thanks to considerable lobbying, provides a useful framework and outlines an entirely new financial support package based on public goods.[1] The devolved nations have responsibility for designing their own support schemes.

The new agricultural transition is revolutionary, though it has taken seven painful years in development. It outlines the move from the CAP structured policy of support, largely for managing land, to one based on payments, mainly for delivering public goods, such as protecting nature, reducing pesticide impact, critical climate mitigation, public access and clean water.

In England this takes the form of the new Environmental Land Management (ELM) scheme, which has three main strands: (1) the Sustainable Farming Incentive (SFI), a set of over 100 good and useful new actions combined with (2) local nature recovery within the existing CS scheme and (3) landscape recovery involving large partnerships working across landscapes or catchment areas.

The huge variety of options to choose from is exciting, ranging across restoring soils and creating river buffers, to agroforestry and hedgerow creation. The approach is piecemeal, offering flexibility – farmers can pick and choose what looks achievable on their land. However, for organic farmers and pioneers already doing deeply agroecological farming on all their land, the ELM scheme may not be so great. Modelling shows that agroecological farming, like organic, can deliver the food and environmental outcomes needed, including climate mitigation, alongside other shifts in consumption.[2] So far, the government has not recognised this nor facilitated farmers to see the whole land and its assets as tools for business and public-good delivery.

Environmental experts and parts of the farming community have welcomed this new approach, albeit with reservations about the design, payment levels, complexity and scope. There remains an enormous gap in terms of advice, training and demonstration for farmers who are new to the idea of public goods delivery. The whole package lacks support for new entrants into farming, for advice provision, for sustaining the agricultural workforce and for encouraging food production on peri-urban land.

Finally, it is worth noting the recent and continuing harsh market environment, with the price of land itself and inputs like oil and chemicals growing massively. Will the new schemes help to ensure farmers can survive and deliver these vital public goods? There is a real need to make sure there are safety nets and adjustments, as the new schemes are rolled out, so we don't see bankruptcies, farm amalgamations, job losses and environmental harm.

Post-Brexit issues

Brexit had other huge impacts on food and farming, including new trade deals, changes to border arrangements and checks (especially with the rest of Europe), and access to workers and seasonal labour.

Many have serious concerns about the new trade deals that have and are being negotiated. Throughout the debates on Brexit, lobbying by MPs of the small yet influential European Research Group, and others, called for cheaper foodstuffs from wherever we could get them. Yet the UK already has some of the cheapest food in Europe. To address issues of access to food, we should target incomes, wages, house prices and welfare, rather than make food ever cheaper in a race to the bottom on standards, which the public does not want.[3] UK farmers foresee major threats to their businesses and impediments to their efforts to help nature and reduce climate impacts if they have to compete with imports of cheaper, lower-standard produce,[4] such as those permitted under the 2023 UK–Australia free trade agreement.

Public health is a major concern too. For instance, the Toxic Trade reports on pesticides and on antibiotic use in livestock showed major differences in regulations in potential new trade partners.[5] The US allows many more pesticides than are authorised for use here and permits higher residue levels. Government plans to tackle obesity could also be at risk. The Sustain Alliance's report, Trick or Trade,[6] revealed how demands from future trade partners could lead to a flood of cheap, unhealthy foods into the UK and could threaten plans to halve childhood obesity by 2030 by challenging policies to promote healthier eating. Breaking with the EU has already been hard, but if we diverge in food standards – such as on gene editing, and pesticide or antibiotic use – our farmers and food manufacturers are likely to find our biggest, closest and most amenable marketplace quickly closing its doors.

Another major headache has been increased friction at our borders, and this is becoming a real problem affecting supply chains and prices. Paperwork, hygiene control, tariff and tax checks, delays and freight issues have proved pre-Brexit fears accurate. Food and drink exports overall have dropped because of the complexity and lack of capacity at all parts of the supply chains and border controls is causing buyers to go elsewhere. The EU export market has for a long time been the logical place for UK food exports for which the much-hyped global market, a Brexit 'solution', is really not a replacement. The short-term impact on jobs and livelihoods and long-term structural change will be immense unless we do something to smooth frictions and better match UK demand with supply.

The impact of Brexit on low-income consumers has been severe. The negative effect on food prices and availability has been intensified by COVID-19 and Russia's invasion of Ukraine. The lack of resilience in local food systems is a disaster driving record numbers to voluntary initiatives like food banks.

There is undoubtedly a welcome post-Brexit legislative push to drive actions and budgets in ways that should, in theory, help farmers and the planet. However, few doubt that dealing with the major critical issues that face us – from tackling climate change and nature loss to antibiotic resistance and obesity– will be any easier alone.

113

Organic market garden

Sweetpea, Caxton, Cambridgeshire

ADRIENNE GORDON

SOIL IS ALWAYS ON MY MIND.

WITH THE AGROECOLOGICAL APPROACH IT'S ALL ABOUT FEEDING THE SOIL — THAT'S HOW WE GROW HEALTHY PLANTS.

FARM PROFILE

YOUR NAME(S)
Adrienne Gordon

REGION
Cambridgeshire

SIZE *(land area owned/managed)*
4 acres (1.6 ha)

RAINFALL *(average annual)*
560 mm

SOIL
Clay-loam

ANNUAL AVERAGE TURNOVER
£16,000 (2022);
£32,000 (2023)

LEGAL FORM OF OWNERSHIP
(company, sole trader, trust etc.)
Sole trader

YOUR APPROACH
(self-defined, as you see it)
Sweetpea Market Garden
aims to support our workers
with a viable income while
providing affordable, accessible
nutrient-dense food for our
local community. We will
provide education and training
opportunities for people looking
to learn more about sustainable
food production.

WEBSITE
sweetpeamarketgarden.co.uk

BUSINESS/FARM NAME
Sweetpea Market Garden

BUSINESS TYPE
(e.g. mixed farm, smallholding, forestry)
Market garden

PRODUCTS *(principle business outputs)*
Horticulture

ALTITUDE
60 m

**YEAR ESTABLISHED AND/OR
TOOK CONTROL/BOUGHT**
2022

EMPLOYEES *(include yourselves)*
1 (2022);
3 (2 part time, 2023)

TENURE
(of land ownership)
Rented

**KEY AGROECOLOGICAL
PRACTICES**
Min-till using hand tools
Planting for pollinators, beneficial insects
and biodiversity
No chemicals (only amendments are
homegrown, seaweed or compost)
In transition to organic certification
Keeping the soil covered (preferably with
living roots)
Sharing knowledge – through volunteer
days and with other growers in the
Landworkers' Alliance
Paid work – an aspiration personally as I
can't support myself yet but our workers
are paid a living wage

I founded Sweetpea Market Garden in 2022; *Six Inches of Soil* followed me from day 1. I grow on a 4 acre (1.6 ha) newly established market garden site in Caxton, Cambridgeshire. The land is part of my landlord's 470 ha arable farm.

WWOOFing my way to Sweetpea Market Garden

I've been passionate about good food since my early 20s. After working as a barista in Cambridge for several years, I decided to travel to look at how people were growing food. I got interested in permaculture and set off to New Zealand and WWOOFed for community gardens, permaculture projects and market gardens.[1] I landed a traineeship at Pakaraka Permaculture, a quarter-acre market garden. Returning to the UK, it was difficult to find work locally, and I ended up in Sussex, at Barcombe Nurseries in 2020, working on a veg box scheme that's been going for over 30 years. Barcombe is a great example of how clay soil can become really productive if managed well. The following year I was doing maternity cover at Peapod Veg, managing a 130-share, 4 acre (1.6 ha) community supported agriculture (CSA) project in Hastings, Sussex. But I wanted to be back in Cambridgeshire.

Getting access to land is really hard. You need money, family backing or community support, or all three, to access land and start a new and successful business. It's almost impossible for new entrants who don't have this. Cambridgeshire County Council is one of the dwindling number of councils that still operate their own council farms. I looked into that option, but the tenancies are for much bigger areas of land than I could use and are hard to get.

Initially, I had no idea how to go about securing land. I looked for somewhere around my parents' home and thought about how I might fund that. I went on some courses that taught how to start a horticulture business, run by organisations like the Organic Growers Alliance, and I sought advice on how to find land. Connecting with other people doing the same, led me to volunteer at conferences, such as the Oxford Real Farming Conference (ORFC), because I couldn't afford to go otherwise. A post I left on the 2021 ORFC noticeboard, seeking land in Cambridgeshire, was seen by Tom Pearson (now my landlord, see farm profile, pages 294–303) and we connected. I signed the lease for the 4 acre (1.6 ha) plot in October 2021, having been in discussions with Tom since January of that year. He was the only person who got back to me actually, and he just happened to have land that's really close to where I grew up. It felt really serendipitous to return home.

> **TOM:** It was quite exciting to see the noticeboard post, and we got talking online, did an interview and took things from there. The idea had been mulling around in my head for some time. I was interested in creating more product variety on the farm. We wanted to build a local community into which we could supply. That can be a little bit difficult on combinable crops, such as wheat and barley that need processing, so we wanted to build up a customer base using fruit and vegetables, which was why we started looking for a market gardener to come on site. We have a long-term plan to introduce education to the farm, and market gardens are far more exciting places to hang out than fields of wheat!

I don't have a family connection to farming or growing, and it wasn't a career path that I ever imagined would be available to me. I definitely grew up with a lot of middle-class privilege and went to university not really sure what I wanted to do but knowing that I cared about nature and climate change. I wanted to do something that had a positive impact on the world.

I've always loved food. I've been vegetarian since I was eight. I learned to cook when I left home and always loved it. I've worked in cafes a fair bit and thought about being a chef but knew it wouldn't suit me because of the late nights and long hours. And so, I decided to do farming instead, which is early mornings and, erm, longer hours. I couldn't imagine doing anything else now.

I noticed that there was a real lack of local food in this area. Also there weren't many places doing it how I wanted to do it, which is very low impact using mostly hand tools, without being highly mechanised with big tractors. I often felt overlooked when applying for jobs. For instance, a manager for a market garden is often expected to be a handy**man**. Leaving gender aside, you need a really diverse skill set. Not only do you need to look after crops, you also need to work machinery, you need to fix machinery, you need to bodge solutions, or you need to be a plumber and sort the irrigation out. That's a really steep learning curve. Doing that for someone else is quite hard, especially in an industry where the margins are so small that people don't want to pay for you to be spending time learning. But doing it on my own means I can give myself the time to learn, and I can do it the way that I want to, without being dictated to by other people.

Plan for year 1

In March 2022 I had a new polytunnel, lots of seedings and a bed of garlic we put in the previous October, just after signing the lease. The plan for year 1 was to focus on salad crops mainly, which is one of the more profitable things that you can grow in this country. In the polytunnel were tomatoes, cucumbers and peppers, and outside there was a high-rotation, succession-planted cropping area with plants in the ground for a short period of time – as soon as one crop was harvested, another went in – making the most of a small amount of space.

In terms of actual growing space, I started on half an acre of the 4 acre (1.6 ha) plot. The whole thing was too big for me initially. Tom drilled a cover crop into the rest to keep the ground covered and have roots in the soil.

My intention from the outset was to certify as organic. The principles of not using any chemical fertilisers or pesticides, and for the whole

production process to have the lowest environmental impact, is very important to me. I'm very lucky because Tom has been transitioning to regenerative farming, so alongside not tilling this plot for a few years, he's also cut out the use of pesticides and insecticides. He's not organic though, so the land wasn't certified as organic when I arrived. The organic conversion process takes 2 years.[2]

I initially targeted markets to reach customers and build a brand. This farm is ideally suited in the middle of three historic market towns, which is amazing. I wanted to be in the local markets at St Neots, St Ives and Cambridge. At the time, St Neots and St Ives markets were both fully subscribed and not taking on new people, but I secured a weekly stall in Cambridge. The farm is also right next door to Camborne, which is a new housing development, home to 10,000 people.

In addition, I sold to the Cambridge Food Hub, which is an innovative scheme that helps support local food businesses in the area. An option taken by many growers is to set up a CSA scheme.[3] It's basically a structure through which people commit to supporting a farm by helping them with the upfront costs for the season and then committing to purchase the produce. It provides some cashflow certainty. Most operate on a share basis, with supporters paying weekly or monthly for a share of the produce. I definitely like the CSA model. It builds community and connects directly with customers locally. Starting out, it's quite difficult to manage because you need to plan enough diversity of cropping so that people aren't eating the same thing all the time, and you need to be confident about having enough ready each week. I decided to look at this for year 2.

I prepared a detailed crop plan in a spreadsheet. It had the date, week, crop and variety. Then I set up formulas to calculate the number of seedings I would need for the number of beds – taking into account the length of the beds and plant spacing, to ensure I had the right amount for

NOT USING ANY CHEMICAL FERTILISERS OR PESTICIDES, AND FOR THE WHOLE PRODUCTION PROCESS TO HAVE THE LOWEST ENVIRONMENTAL IMPACT, IS VERY IMPORTANT TO ME.

Cambridge Food Hub

The Hub[4] is a local organisation that was set up to increase the accessibility of sustainable food while supporting local producers and small businesses. It provides producers with routes to market. The idea is that collaboration reduces overheads and waste for each individual business. It's managed online. I list the products and set stock levels a week ahead, and then harvest to order. They collect from me and consolidate the deliveries, which reduces the number of vehicles on the road and is really invaluable. Prices are wholesale, but in return I have consistent business and access to local buyers who are interested in local food, so some of the work is done for me.

planting out. I had a similar spreadsheet for direct sowing (pictured). So, for instance, among the things I needed to sow in week 12, starting 21 March, was 225 beetroot plants.

Ideally this would all be fixed each year in January, but it has been changing as I figured out my market and how much I needed to grow. As I got a better idea of how well crops grew on this land that was new to me, I adjusted the actual amounts. Some produce had weekly showings throughout the year, such as lettuce and other salads. At the end of the year, I reviewed performance against the plan and took that into account in planning for year 2.

Transplanting plan.

I have wanted to do this for so long. It's been a really big dream of mine that I was so excited about starting, but what I didn't envisage was the kind of fear–freeze paralysis that came with actually starting such a massive project. It was just me; I had friends and family helping, but this was going to be my thing, and it was, and continues to be, scary to think about putting my business and me out into the world, not knowing if people will want to buy things from me.

Regenerating the land

Tom separated off the land and sowed a cover crop in September 2021. It's highly productive land. Cambridge is well known for having good farming land which is used mostly for crops rather than animals. A cover crop is a mixture of different plants that are sown to bring fertility and improve soil structure. It's really bad to leave soil bare, especially over the winter, because it exposes it to erosion and loss of nutrients. The plants help protect the soil. Tom sowed it as a way of improving the soil before I moved onto the plot. We paid a lot of attention to restoring the main section of land that I was not using in year 1 so that in year 2 onwards it would be really healthy, vibrant soil. It was subsoiled because historically it was turning headland for the larger field it was part of, so it was very compacted. After the subsoiling, a second cover crop went in.

In April 2022, before the subsoiling, a local shepherd called Jacob, who runs a mobile flock, grazing for various landowners including solar farms and some National Trust properties, brought his 80 sheep in to graze-off the first cover crop. We could have just cut it, but the benefit of the sheep is they mow it down effectively and poo everywhere; we also get the benefit of the gentle trampling down of the manure which will add nutrients and organic matter, which is food for the worms and other microfauna in the soil, which then improve the soil structure and release locked-up nutrients.

Jacob said it had a lot of great feed on it and came at a great time for him as it's unusual for people to still have a lot of good-quality grazing available in March and April. Such connections develop community that brings further benefits. When chatting with Jacob, he told me that he had unsold wool that's been standing for about 2 years in his yard. He offered it to me, and I'm hoping to use it to make compost or mulch.

Going forward, I will probably always have an area in my rotation that's sown for a green manure, so I'm really happy to have found someone to help with grazing. Taking up livestock as another operation by yourself is a big commitment. It's a different skillset. We each have our specialities, so working together is much more resilient than one person trying to do everything themselves. Agroecological systems are about the whole complex picture and they are about the people involved in them as well. It's important how we interact with our environment and those working alongside us; if we give to the system, the system and others within it give back to us.

Soil is always on my mind

Cover crops have an important role to play in retaining the moisture in the soil, which is always damper and more workable under dense plant growth than in the uncovered areas, which dry out and are more clumpy. Generally the clay soil here does a good job of staying wet apart from at the very surface where it can dry off, crack and form a pan, which is problematic – it can be hard to break up the surface.

Soil is always on my mind. Along with what the weather is doing to it, especially on clay. If it dries out too much, it's unworkable. If it's too wet, it's unworkable. There's a constant managing of the soil in order to be able to plant into it, which I didn't even consider when I was growing plants in other places. This was quite new to me; I was really lucky to have experience working with clay soil last year because I don't think I would have known what I was doing now without that.

Clovers, part of most cover crop mixes, are brilliant for what they bring to the soil. They are a legume and nitrogen fixing. They've got really nice deep roots which get down into the soil, and on those roots are little nodules, little lumps, which fix nitrogen: bringing down nitrogen from the air, through the plant and into the soil.

The agroecological approach to soil is all about feeding the soil, that's how we grow healthy plants. So rather than thinking about giving directly to the plant, we need to focus on adding things to the soil. That means that the soil builds its fertility and unlocks the things that are already in it.

IF WE GIVE TO THE SYSTEM, THE SYSTEM AND OTHERS WITHIN IT GIVE BACK TO US.

Clover root with nitrogen nodules visible.

I love that in clay soil you find these little holes where there's a worm just curled up inside; it can be the densest clay, and you break it open and there's just a little ball of worm. I don't really like using human terms, but they're the engineers of the soil. They do a lot of the work of moving nutrients around and breaking up the soil. They eat plant matter and then turn that into soil. They are the farmers really.

Fourteen-hour days

During the first few months, I was doing farm-related things for about 14 hours of the day. I woke up and wrote emails for a couple of hours, and then I got the bus to the farm. I was at the farm for 9 to 10 hours, and then I went home and researched packaging options or ordered samples or did any number of admin things that I had intended to work on over the winter but didn't get done. It was a lot of work. But that is the nature of starting a business. I think anyone starting any kind of business would say that they work a lot.

Developing the brand, the presentation of Sweetpea Market Garden took a lot of time, but I really enjoyed it. There's an element of creativity involved and excitement about the story that I'm telling and how I'm presenting it to the world. Part of that was the packaging. Until the organic conversion was complete, I couldn't advertise my produce as organic. But actually, organic is not always the best way of communicating what Sweetpea is doing. So I settled on 'planet first'. I was trying to come up with a way of describing all our ethics in a really short, pithy way and 'planet first' just seemed to work.

Getting the labels and packaging right took a lot of thought. An illustrator friend did the logo that is used everywhere – including on the labels (pictured), which also prominently feature the location, Caxton Cambridge CB23, as I really wanted to emphasise the fact that it's local,

Sweetpea pot label.

seasonal and embodies minimal food miles. There is space for the product description and brief instructions to keep refrigerated and wash before use because the only things I'm going to be labelling are things like salad that need to be stored in something airtight. And then 'Planet First Produce / Grown with love, without chemical inputs, using agroecological principles' and then the website at the bottom.

The pots look like plastic but are compostable. I needed them to package microgreens because they don't keep or transport well in a bag. I had to get samples to work out how much each would hold and therefore how much I would charge.

There were two different styles: one with a foldable lid and one that's got a removable lid. I preferred the removable one in terms of ease of use when loading produce into them. But actually putting the lid on is a little bit tricky and that is something that you really have to consider – when you're trying to pack 50 punnets of microgreens, you want it to be the most efficient way that you can because the harvesting and the packing ends up taking most of a market gardener's time.

Packaging is a minefield in terms of sustainability; a lot of small-scale fruit and veg producers just can't manage to pay the extra for compostable packaging and still charge affordable rates for their produce. It will be something that I'll have to keep an eye on; but I really wanted to start as I meant to go on and to have compostable packaging. I use as little as possible. Most of my produce going direct to customers at markets or to cafes and restaurants doesn't need packaging, like bunches of chard and kale just tied with a rubber band. The only things that will be packed are the salads and microgreens that just do not store for long enough without packaging.

I will promote returnable options, particularly for repeat customers. The pots are not inherently good just because they're compostable. There's still carbon involved in the process. It still takes energy to make them, and the plants have to be grown somewhere to provide the materials. It says commercially compostable, so they probably need an intensive composting method to actually break down relatively quickly; they don't always break down that well in a garden compost.

A farm education

On 10 May Tom and I were delighted to be invited to FarmED (see page 272), the award-winning demonstration farm and education centre in the Cotswolds set up by Ian Wilkinson. Ian took us through his personal journey; he explained why he set up FarmED and what it did. While on a farm walk, we discussed many of the issues affecting the agri-food system, and the responses that FarmED was promoting across the site. Of most interest to me was the visit to their CSA and talking with Emma Mills the market gardener at FarmED.

IAN: We have a CSA scheme. When they started here, they were on about 1.5 acres (0.6 ha) and they've now moved to 5 acres (2 ha) because the customer base has expanded: to 127 shares, plus 50 shares for FarmED's kitchen. We, as the farm, rent the land to this social enterprise for a small peppercorn sum. The food moves from here to people within 5 to 10 miles (8–16 km), no more. It's collected every Friday from here and also from the local school. It's a very short supply chain.

There are five people employed, who make a living wage. Some are nearly full time and some are part time. There are

Ian Wilkinson (left) with Tom and Adrienne in a FarmED wheat field.

also volunteers that come here too. They're not essential to the CSA's viability, but they help and they enjoy the space. So, from a well-being point of view, it's wonderful. And the food they produce is amazing. The CSA is not certified as organic, but everybody knows the provenance. Everybody understands how it's grown. They have wonderful communication with their supporters, and the supporters pledge to support for a year.

There's a long waiting list for people wanting to subscribe to this type of scheme here. And in this case, it's not expensive. The CSA is £30 a month, and you get a share every week. From our farm's point of view, it's great because we have people coming down the farm drive every week to collect vegetables, which means the other micro-businesses here on the farm all have a direct route to market without having to have a marketing budget.

ADRIENNE: It's really heartening to hear that they are actually able to pay a living wage. So many of the businesses that I've looked at, on my journey of learning how to grow food, either rely heavily on volunteer labour or on external grants or funding. Is it all year round?

IAN: Yes, but there's the hunger gap right now. Although it feels like spring's here, as you know, there's no food at this time of year from a horticultural system, or very little. Which is why the polytunnels are really important. That's the restricting factor on this particular CSA, the amount of polytunnels they have.

ADRIENNE: Do they buy in any produce?

IAN: No. There's nothing coming in to fill out the boxes. Just what is grown here.

ADRIENNE: So, with the hunger gap, people are getting less produce in their share?

IAN: Exactly. They get a share of everything that's produced at the time. Right now, there might only be three or four items. But for the rest of the year there's much, much more.

The tunnels were funded by local social projects: a solar farm and a wind farm. They enable us, as you know, to have a much wider season of production.

ADRIENNE: I wonder if the local funding for these polytunnels was helped by the fact that they were starting here and had the reputation of FarmED to build on.

IAN: Actually, the CSA was running before it came to the farm. A lot of people wanted to subscribe, but they had no more land where they were. So they came here because we had land available, and we were very keen for people to come because it really has opened up the farm. This was the first new enterprise on the farm that was not controlled by us as the farmers, and it really opened my eyes as to how productive it is, how it brings people to the farm and has reinvented the farmyard. Our whole enterprise is evolving around the micro-businesses on the farm and an educational centre.

ADRIENNE: We need this to be replicated for every village, for every town.

IAN: And you can because it's not that competitive. There are only 180 shares a week, but there are thousands of people living close by, even in a rural area like this. There is so much scope. The kitchen-garden people here don't want to increase the number of people they're supplying veg too. They want to train the next generation; they want apprentices in here, and then hopefully they will go on, in years to come, to create brand new CSAs or other models elsewhere – it's a different kind of expansion.

ADRIENNE: It's so useful for me to see this because it shows me how I might go about starting a CSA myself. I definitely would like to talk to Emma about what I might do next year.

ADRIENNE: How did you get into growing, Emma?

EMMA: It started around 2016. I'd been going to the ORFC to try to learn a bit more about diversifying farms. My folks are farmers in Wales, and we've got Welsh Blacks and Welsh Mountain sheep, but I kept finding myself in these lectures at the ORFC about market gardening, and I came out of them during one of the coffee breaks and said to a friend of mine, 'We don't really have anything like this around us. We don't have community supported agriculture.' We've got lots of organic produce being

129

grown, but it seemed to be a luxury commodity, and I felt access to good food shouldn't really be limited in that way.

So that was the inspiration behind it. My friend had been offered a tiny patch of land, one-fifth of an acre, about 3 years previously. She went back to that person, the land was still available, and then we had no excuse to not do it! That's basically how I got into it, but I think the driving force was definitely a deep-rooted sense that I wanted to work alongside nature. Conservation has been a really big part of my family, growing up, and I think it just stays with you, and you feel like what you do in life needs to align with how you feel inwardly.

I didn't have masses of experience, but I work with an amazing team that I developed over time. One of whom, Dan, had already been growing for about 3 years on a much bigger scale, which gave us the confidence and impetus to scale up over time.

ADRIENNE: Starting small is definitely easier. I've got 4 acres (1.6 ha) but am only using about half an acre at the moment.

EMMA: If I'd been in a situation where I had all 5 acres (2 ha), I think it would have been a bit overwhelming. Starting with one-fifth of an acre meant that we were able to start with just 30 boxes a week. It meant that we had really strong relationships with all of those customers. And what I didn't envisage happening, when we came to expand, was that those core pioneer customers ended up being an unpaid marketing department.

ADRIENNE: Did you have Dan on board from the very start?

EMMA: Pretty much. It was the two of us at the beginning, and then he came in during the first year. We started our first share in March 2017. We called it a salad share as we knew we could produce that every week, and the salad was complemented by other vegetables that were in season at the same time. That took the pressure off having to fill a whole veg box right at the start. We still call it a salad share, but it's a vegetable box share.

ADRIENNE: That's one of the reasons I haven't gone down the CSA route this year, although I've got a lot of growing experience, I've never actually had to do the crop planning. I wasn't confident that I would be able to produce a regular amount all year around.

EMMA: For many, I think it's stepping into the unknown because it's such a different consumer concept – the idea that you're going to basically be sharing the risks and the benefits of farming with the customer. The risk is something the customer doesn't really know about when they buy from a supermarket. It's a story that hasn't been told. I suggest starting a CSA as a micropart of your business. Be really transparent with your initial customers, then they will trust you on the journey that you're taking together.

ADRIENNE: How did you go about finding your first 30 customers when you started?

EMMA: Because it was just 30 and there were three of us on the team, it meant it was really easy. We reached out to friends first of all, and we also created collection points that were quite close to primary schools. A couple of us had younger kids so that made sense. We wanted collection points that were at places people had to come to anyway so that the food miles were low, but unintentionally that meant that we also had a really wide network of people to market to.

ADRIENNE: In terms of income, when you started out did you have to work without paying yourselves, as I am doing?

EMMA: We did. I think that can also be a bit of a barrier. With a CSA your income starts at the beginning of the season. But to have any produce by that point, obviously you've had to start planting and growing months before that. So that first year is hard.

ADRIENNE: Something that's really important to me, and it's definitely not being achieved yet, is earning a viable income for me and future staff. Have you managed to do that?

IT'S SUCH A DIFFERENT CONSUMER CONCEPT —
SHARING THE RISKS AND THE BENEFITS OF FARMING
WITH THE COMSUMER.

EMMA: Yes. We are on £10–15 an hour, so it's not an amazing wage, but having said that, we also have really flexible working hours. I'm a single mum with three young children and for me it's really important to be able to take them to school and be there to pick them up. I can do that working here.

ADRIENNE: In terms of long-term sustainability, you have a strong team. That's really important and something that I'm hoping to build up to.

EMMA: I think that's it, having a team means you can help each other. So it's not always the same person shouldering the burden. Also, you get to benefit from a range of interests and skills. One of the team, Matt, is really interested in creating new little inventions like a low-tech way of automatically opening up the polytunnels and closing them. Engaging with our local community and communications in general come easily to me. I think working as a team within the wider community of growers can really help too.

ADRIENNE: I don't think I would have the confidence to do that if it wasn't for the Landworkers' Alliance.

EMMA: There's no way I would have done this on my own.

ADRIENNE: We can see that it works, and you can make an income from growing veg for a community and sell it at an affordable price as well. Why do you think it is that we don't already have a CSA everywhere?

EMMA: I feel really optimistic actually. There's going to have to be shifts at policy level. But there are also a lot of individuals out there who want to know what they can do to take an active role in bringing about change. And this is something that people can literally do tomorrow, by choosing to buy locally. I get asked all the time if I'll expand the business, and I genuinely feel that what we're doing now, 180 boxes a week, is enough for here. But there's no reason that 5 minutes down the road there couldn't be another CSA doing a similar thing.

SOMETHING THAT PEOPLE CAN LITERALLY DO TOMORROW, TAKING AN ACTIVE ROLE BY CHOOSING TO BUY LOCALLY.

When it comes to farms and farmers wanting to create direct ways to market, this is a great way of doing it. If they offer a grower 5 acres (2 ha) of land, they can build a community of over 100 people coming to collect veg boxes, and it builds. We can see it happening already here. We work with a micro-dairy, so people collect their veg box and their milk, we work with a baker down the road who brings sourdough bread on a Friday. We work with Rosamund Young from Kites Nest farm; she brings her organic meat once a month.

I think it's going to take a combination of actions. Thankfully, we've got organisations like the Landworkers' Alliance who are doing great work on our behalf at the policy level. As growers and farmers, we can encourage and enable our local communities, but I don't think it's going to be done through what I call 'guilt talk'. You're never going to get anything done in this world if you make people feel bad for the choices they're *not* making. We actually have an opportunity to celebrate the choices people *are* making and enable them to be able to make them.

ADRIENNE: Amazing. Thank you.

I didn't know what to expect before we visited because we were coming to meet Ian, and he's not a veg grower. I knew there was a CSA here, but I wasn't sure if we would get to meet the manager of that. Just being here on a farm that's education focused and that has beautiful buildings and a lovely cafe, and receiving a lunch cooked from food produced here, is a reinforcement of all the reasons why I want to be doing this.

I think it's a really interesting model. I think it's very positive and aspirational. I think this is how we are going to make change. As Emma said, you don't get people to do things by making them feel guilty. You show them the positive solutions and then they want to be a part of it.

At the market

In June and July 2022 we started attending markets regularly. We were pretty much saying yes to any local markets that we could get into because most of the big established ones had a veg trader already, and they didn't want more than one. So, if we were asked, we went. I enjoy the market stalls. I love being able to talk to the customers directly. That's my favourite bit about it. It was quite stressful finding markets and getting all of the paperwork sorted. We got a lot of rejections at the start, and without a lot of produce, it was really hard to sell ourselves.

Ahead of each market, I made a provisional picking list based on what I expected to be ready, which was then adjusted as we packed – based on what was actually ready. This ensured we had an accurate record of what we took with us to market. We use a point of sale (POS) system to take card payments, which keeps a record of everything sold.

The picking list also confirmed the units (e.g. by weight, bunch or head), the sale price per unit and the value per product type. Totalling this up gave a maximum sale value and an indication of what we might take. For instance, the picking list for the Fulbourn market on 18 June (pictured) indicated we could take up to £230. For the start of the business that's OK, but really a minimum of £500 per market was needed to make it worthwhile doing them, given the time that was involved.

Our income was uncertain. We couldn't predict how we were going to do at the markets. We could calculate from my spreadsheets how much

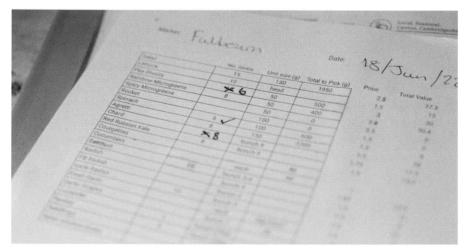

Market picking list.

we might make if we sold everything. But it depended on whether people bought, and that varied between markets and between days at the same market.

We took some of our spare seedlings, when we grew more than we needed (or lost fewer than allowed for) to a few of the little markets, and they did really well. So, we may in future grow more to sell.

By 20 July we'd been to four markets in Cambridge. It's 10 miles (16 km) away from the farm so not too far and a historic market in a busy city. I was hopeful that the Cambridge market would perform well,

Beetroots

Beetroots are one of my favourite things. I love harvesting them because, when you pull them out of the ground, you can smell the earth, and it's just such a happy smell. They store well; they're really easy and quick to harvest; there's no effort involved in making them look nice; they come readily presented in a beautiful package: their skin; and they look lovely with their fresh leaves still attached, which you can also cook when fresh. Our ones are wonderful grated raw in salad as they lack the bitterness of the supermarket ones, and roasting them is great too, you just pop them on the barbecue, wrapped in foil. My dad started putting them in spaghetti bolognaise alongside carrot. And there are so many wonderful varieties, such as Chioggia, all pink and stripy inside.

but we found it's definitely not what it used to be – maybe it didn't help that it was Wednesdays, and people weren't used to doing their vegetable shopping at the market that day. Our position wasn't great; the footfall in that bit was low, and we were kind of hidden around the back, so people had to find us.

But when they did find us, they came back. We quickly got some really loyal customers who came to every single market. Talking to other stall-holders, they said it could take a year or two to build up a customer base. The St Ives market, that we later got in to, is a farmer's market, and there are more people who are doing their veg shop. Whereas in Cambridge the market has become less of a food market and more services, like bike repairs and stalls targeting the tourists and lots of hot food, but not so much produce. We do have a really good neighbour who sells bread, and he recommended us to all of his customers.

I'm told the Cambridge market really suffered during the COVID lockdowns – in terms of traders retiring and not coming back, but also erosion of the customer base. Generally, I think there has been an increasing movement towards delivery. I don't think Cambridge is going to be a long-term outlet for us. I think places where people actually live, rather than in the city centre, could be more viable. The satellite towns and the suburbs might actually be a better market for us. There are likely to be more markets starting up in the area, in the villages closer to the farm. The local district council, South Cambs, made funding available to support the setting up of new farmers' markets, hopefully we're on the cusp of things happening.

Quiet day at Cambridge market.

Six weeks, no rain

By early September we'd had over a month and a half without any noticeable rain. This area of the country does have very low rainfall, so dry spells are expected. But it was really extreme in 2022. It was quite difficult to establish new plantings, and the pest pressure was higher. One of the biggest pests on the farm site is flea beetle, which is particularly bad in this area of the country because there's a lot of oilseed rape grown – a nursery for flea beetle. They also really thrive in hot temperatures so were worse than normal.

What they like to do is eat the leaves of any brassica: rocket, radishes, things like that. I actually had it in my crop plan not to grow these things during the hot weather, but what I didn't expect was that the beetle would also impact things like kale because in other places that I've grown it they just haven't been a problem.

We have a whole-site irrigation system, which I put in because of my experience working in other places where we were constantly moving water pipes around. So here I wanted to avoid that. I invested over £1,000 in plastic piping. There's a main pipe that goes all the way down the main cropping area and every bed has a spur with a tap that I can turn on and off that feeds drip hoses. This system reduces the amount of water loss from evaporation. If you water the soil surface, the water goes straight down into the soil. It encourages the plants to root deeper, in contrast to sprinkler systems that water the leaves and can create very shallow-rooted plants.

Even this groundwater system wasn't enough for us to weather the drought. It's not just about volume; rain has a very different effect because it's better distributed over the site, and it's got different minerals. It brings life with it, whereas mains water, which is chlorinated, is not as good.

There's no way of not suffering from almost two months with no rain. I felt really terrible by the middle of August because I could feel how dry everything was; the soil had big cracks, and it made me feel like a bad land custodian. It was really upsetting to see it like that.

THE SOIL HAD BIG CRACKS, AND IT MADE ME FEEL LIKE A BAD LAND CUSTODIAN.

It was very hard underneath too because it dried up all the way down. That was why all of our new seedlings struggled. The only things that grew were thistles and other deep-rooted plants which were able to get down and access the moisture that is deeper underground. Anything shallow rooted, most of the weeds but also our new crops, really struggled to get going. The plants that were already established, like the courgettes and chard, were OK, but very few of our winter plantings were established. Our squash yield was greatly reduced, from three or four fruits per plant to one, and we didn't have winter crops like cabbage or kale in. I had to take the decision to finish my sales at the end of November because I just didn't have the produce to sell into the winter, and to look for another job until the new season.

It's not surprising. We have known for a long time that our climate is becoming more unstable and uncertain and temperatures have been rising. That's why I think it's really important that we have local food production and that we improve our soils so that they can cope with variable and tougher conditions – it's essential. Applying mulch and compost increases the ability of the soil to retain moisture, especially with clay soils. If you think about clay, how it behaves, it dries out and cracks. But if you add enough organic material, and it's incorporated, it doesn't dry out in

the same way because the organic matter breaks up the clay structure and doesn't allow it to coagulate in the same way. The organic matter also holds more water, like a sponge.

One of my failures this year was that I wasn't able to establish any cover crops, which is one of the ways to maintain and improve good soil health. Roots hold the soil together, maintaining and feeding healthy soil biology. They create pathways, so the water can actually get into the soil much better than where there are no roots. Dry soil doesn't allow for water to move through it. It's hydrophobic. It actually repels the water. When it hits really dry soil, it just pools on the surface or runs away. But if there are plants, then it will get in.

Next year I'm looking at investing much more in mulching and ground surface cover to try to reduce the amount of water lost through evaporation. I will have straw from Tom, our landlord, and also the wool from Jacob, the shepherd whose sheep were grazing here earlier. I'll make sure that our paths are covered but also try to cover more of the soil in between the crops to keep more water in the ground. The rainwater-harvesting system will be online too, meaning we can irrigate with rainwater, which is so much better.

The Apricot Centre

In September I was lucky enough to spend 2 days with Marina O'Connell at the Apricot Centre on Huxhams Cross Farm near Totnes, Devon (see farm profile, pages 284–289). We had a great discussion followed by a farm walk. On the second day, I joined her on her stall at Totnes market.

ADRIENNE: Can you tell us how you began this journey?

MARINA: So, the journey began before this farm. I've always been a horticulturalist. My first job was at nearby Dartington Hall, where I was the market gardener and trained young people. Then I managed to buy a tiny bit of land in Essex and was growing there for about 20 years. I became aware that the Biodynamic Land Trust was buying this farm and looking for a farm partner. I was keen to move back down to Devon, and so we decided to go into a formal partnership. The trust bought this farm and we became the first tenant in 2015.

ADRIENNE: Can you tell me a bit about how you farm, what your approach to farming is?

MARINA: There are six fields, so about 34 acres or 14 hectares, and it had been farmed by the Dartington Hall farmer of about 40 years. First, we undertook a permaculture design process. Then straight away we put the land into a recovery programme and started the biodynamic conversion process (see Chapter 2). We also used a huge amount of agroforestry all the way through the farm. We combined those three methods: permaculture design and biodynamic processes, with agroforestry woven all the way through it. I quite often call it regenerative farming because it's easier for people to understand.

ADRIENNE: When you started you went through quite a process I assume?

MARINA: It was a bare site that had been industrially farmed for decades. We just went for a swift detox, so to speak. We had all of the arable fields ploughed and sown with deep-rooting green manures. We left the manures in for 2 years, and then slowly brought the fields into production. We're a fairly small farm, but that still cost quite a lot of money. So, we just had to take the financial hit, and it was really quite difficult.

In addition to the deep-rooting manures, we put on biodynamic preparations which inoculate the soil with the soil biome, a little bit like a probiotic drink. And we did something called keyline ploughing. It is like subsoiling, which cuts a slice through the soil that then lets the air in and the water out, but it is done following the land contours. This helps to stop rain running across the surface, and instead it runs down through the soil.

We also kept topping the green manures off. So that meant we were feeding the biome as well. It took 2 years for the soil to recover, but we needed that time anyway to get the farm up and running and sort out the infrastructure.

ADRIENNE: How would that work on a larger farm?

MARINA: If you had a bigger farm, I think that you probably would want to stagger the transition in stages. As you go into that transition, you do stop spending so much money on inputs, nitrate fertilisers and pesticides, so even though you're not making money, you're not spending as much money as you might have been. And if you're not already a mixed farm, you could bring in livestock to graze down the green manures, which is a way of generating some income from that transition period.

Marina O'Connell (left) with Adrienne at the Apricot Centre.

ADRIENNE: So, the costs of transition are quite a barrier?

MARINA: It is something you have to plan and budget for. For many, that will mean doing two jobs or having another income for a couple of years. The cost of the transition period is recognised as an issue. The new Environmental Land Management scheme and Sustainable Farming Incentive should help farmers get through the transition.

The transition process brings more diversity onto the farm – a greater diversity of crops, you might add livestock, you might start growing fruits and vegetables. Usually no farmer has all of those skills, so more people come onto the farm to work, to collaborate and to produce those different crops. They bring different skills, different contacts, they are part of different networks. It's also important to localise your sales – by selling directly to people in and around your region, you also make more money.

When you go down this route you start to regenerate your whole-farm economics: you're not buying expensive inputs and you're selling locally or directly.

ADRIENNE: I'm interested in how you sell your produce. I found that locally there are many farmers markets, but people aren't really attending them. What's your experience and how did you have confidence?

MARINA: With the benefit of doing this kind of work for about the last 30 years, I've seen the market grow steadily. Step by step, it's got bigger and bigger. There's something in just being confident; believing that if I grow this food, I will find a market.

You have to be bold. Be bold and take the next step, and the people will come to you. My experience is people really are desperate for this kind of food. Especially as we come out of COVID, people are really keen to be connected to nature via their food and to embrace seasonality.

It's been extraordinary. I thought the Totnes market would be saturated because this is an epicentre of organic growing around here. We've got Riverford down here. That's the largest organic food producer in Europe. However, I found that cluster-marketing takes effect. People come because they want good-quality organic food. Therefore there can be more growers producing more food because the more there are, the more people come.

We run a veg bag, like a CSA. We deliver food to people. We've got about 150 customers for the bags each week, and we do Totnes market on Fridays as well. We try to engage with our customers – whether through the newsletter or on the markets, in person with our regulars, telling them what's going on at the farm and inviting them in for events.

Cambridge is a very wealthy area, and there's a lot of people, but I realise it's not quite the bubble that Totnes is. With your 4 acres (1.6 ha), I'd say you'd only need about 100 people in a scheme to support you.

ADRIENNE: When you started, how many CSA shares did you do in the first year?

YOU HAVE TO BE BOLD. BE BOLD AND TAKE THE NEXT STEP, AND THE PEOPLE WILL COME TO YOU. MY EXPERIENCE IS PEOPLE REALLY ARE DESPERATE FOR THIS KIND OF FOOD.

MARINA: We had 20. We weren't even growing many vegetables then. We bought them in as we weren't growing enough. We started trading before we were growing produce because we had to make some money. Straightaway, we got some hens to start laying eggs. We connected with the really tiny micro-growers around here and bought in from them. We set up an online shop. We just started. On your scale, I would also go for the high-value crops and target really high-end restaurants.

ADRIENNE: One of the main things that I've struggled with this year is how to communicate with customers in terms of what I will have and what I won't – knowing what I can promise people.

MARINA: So that's really difficult, especially this year with the droughts and the high temperatures. Things didn't behave as they were supposed to behave. Part of the craft of this, basically, is crop walking and crop estimates. You have to do that every week and eventually you get the hang of estimating more accurately what may be ready in, say, 6 weeks' time.

So it's communication, communication and then a bit more communication. We send out emails every single week to our customers, and then on the market we basically talk to people. And I think people kind of get it. It's an important part of the story to say, 'We've had 40 degree temperatures and the lettuce has bolted.' We have to absolutely tell the stories. Because they need to know. So that it becomes a part of their story. Restaurants will also tell their customers, and that's good for their connection and business as well. There's a seasonality, there's a story.

The people who are buying our kind of food, I feel, want to be part of the farm life. They want to buy the stories as well as the food. I believe, as regenerative farmers and growers, part of our job is to tell the story of the farm and what is happening to farming in real time with climate change. That is so different to when you buy food in supermarkets, and there's no personal connection.

ADRIENNE: We've had some trouble breaking into the marketplace. There's been some reluctance from other growers to allow us in.

MARINA: There are a few things I have to say about that. I think we're so used to behaving in a competitive way; we've all been brought up with this understanding that we have to compete, but actually regenerative farming's about collaboration. I've been in that situation, on both sides – I've felt threatened by other people coming in, and then obviously when we moved here, it caused a stir among local growers. But what's happened is everybody has collectively increased their share of the wider market. Everybody's selling more than before we were here. I find that fascinating. Our competition isn't with each other – it's actually with the supermarkets. What we want to do is outcompete the industrial farmers. We have to learn to work with each other. But I appreciate that is difficult when we're all feeling very insecure because we're all on such a financial knife edge.

ADRIENNE: Have you noticed a drop in customers with the cost of living crisis?

MARINA: I was expecting a drop off. We had a dip when the fuel prices went up in April, but we've actually recovered now. I'm slightly surprised in fact. I don't know if it's a Totnes thing that people here are quite happy to invest in us and spend a little bit extra on food.

A fantastic thing I saw in an independent supermarket in Manchester was a chalkboard with a price comparison: the price of organic potatoes that they were selling, alongside how much they would cost in Waitrose, Sainsbury's and others locally. Time consuming I know, but if you did a price comparison on a few products, you probably would find that your food is cheaper than the supermarkets' like-for-like food. We should be making that clear to customers. And because you're part of the circular local economy, money spent with you will benefit the local community more than pounds spent in a chain supermarket.

ADRIENNE: Do you think you've retained the customers because they are connected to your story?

OUR COMPETITION ISN'T WITH EACH OTHER — IT'S ACTUALLY WITH THE SUPERMARKETS.

MARINA: Yes, definitely. Another story you can tell is about the food's carbon footprint: 30% of an individual's carbon footprint comes from food. So, if you make a few changes, you can reduce your carbon footprint considerably – switching to sustainably produced or organic food, switching to local food, cutting out waste, and having two meat-free days a week can cut your carbon emissions by about 20%.[5]

ADRIENNE: Were you able to pay yourself from the start?

MARINA: We had to pay ourselves, to find ways to make money, because there is no magic pot of money. Do you manage to pay yourself?

ADRIENNE: No, I haven't been. But that was my strategy going in. This year I invested my own money to set it up, and I have been earning money, but it's all been going back in to the business.

MARINA: I know that you are on 4 acres (1.6 ha), as I had when I was in Essex. My experience is it's really hard on that size, you can just about make a living for one person. The new ELMs should give you access to a little bit of money. Educational visits help, you get £300 a pop for those, and you can do 25 in a year, so that's about £7,000, and it's always nice to have children on the farm.

ADRIENNE: You also have trainees and teach as well?

MARINA: My very first job was training young people and growing produce with them. It's a high-value skill set. If I run a training course, I can earn a lot more money than as a grower. But I don't enjoy being a full-time lecturer. It's nice to do some training, and it brings a bit more money in, but you need a balance. We've received funding from the Devon Environment Foundation and Devon County Council to run our 1-year level 3 traineeship in Regenerative Land-Based Systems: Food & Farming.

ADRIENNE: Who are the people coming to learn here? What are their backgrounds?

MARINA: It's very mixed. Some are postgraduate, so they've done a degree and are then thinking, 'What am I going to do for a living?' Some have already had a career and are then retraining because they've realised they want to work outside, and some are school leavers. We go from 18-year-olds right through. There are some connecting themes, such as climate anxiety and wanting to do something positive in response to that. And there are young people who had no idea that you can actually have a career farming or growing. What about your background?

ADRIENNE: I learned by volunteering for other people. Then I eventually got a job on a farm. Often the route in is to do a traineeship or to volunteer, unpaid or very poorly paid. I worry that's a barrier to entry: it's just not affordable for many who would like to work on the land. How do you feel about that in terms of equity and access?

MARINA: That's a very good point. I feel regenerative farming and growing really needs to become more professionalised, to be recognised as such and to be paid properly. Why should we be working for next to nothing?

If you're a school leaver now and you want to become an electrician or learn another skilled trade, you do an apprenticeship, splitting your time between paid work and college. The money is so poor in farming that even if trainees were just paid the minimum wage while they were training, most farmers and growers would find that too expensive to be viable. As we all know, if you get an electrician in you would be paying them £200–300 or more a day. If we could pay ourselves, as growers, the same, then we could afford to pay trainees.

ADRIENNE: How much do you think food would have to cost in order for farmers to pay themselves as the electrician does?

MARINA: Good question, I don't know the answer to that. But there are some very interesting statistics, Tim Lang's work showed that in the 1950s we spent something like 40% or 50% of our income on our food, and now it is less than 10%.[6]

REGENERATIVE FARMING AND GROWING REALLY NEEDS
TO BECOME MORE PROFESSIONALISED, TO BE RECOGNISED
AS SUCH AND TO BE PAID PROPERLY. WHY SHOULD WE
BE WORKING FOR NEXT TO NOTHING?

ADRIENNE: Part of my journey, my reason for coming into this work, is I've struggled quite a lot with my mental health, and I've found that being outside and being with nature is really beneficial. How have you found it?

MARINA: Farming in general, it is often said, has the second-highest suicide rate after dentistry. My masters' dissertation questioned why, when we know that being in nature is good for our mental health, does farming have such a high suicide rate.

I looked at different types of farming, and obviously regenerative farms have got a lot more nature in them, and there are a lot more people working on them. So I think they're more regenerative for our own psyches as well. But generally, when you're farming, there are a lot of things out of your control, for instance, the weather, and there's a lot of debt pressure, which then of course becomes very stressful. Personally, I've learned to worry about what I can control. And if I can't control something, then it's best not to worry about it. A couple of my sayings are: 'There's always next year' and 'Two out of three is not bad.'

ADRIENNE: How do you find it to be a woman in farming, and do you think things have changed?

MARINA: I think things have changed hugely, yes. There was a moment at a Growing Communities[7] farmers' market one year, when I looked around and I realised that at least 50% of the farmers and the stallholders were women. And when you go to the conferences, there will be at least 50% women, if not more now. We're hosting ORFC in the Field in 2023, and all the main workshop speakers are women.

ADRIENNE: Is that reflected in industrial farming as well?

MARINA: I'm not that involved in the industrial world, but I think more farmers' daughters are welcomed into the fold now. I think a lot of barriers have dropped there as well – not only more women, but more people of colour and gender diversity.

ADRIENNE: You personally are providing a lot of momentum to the regenerative transition. So tell me about the revolution?

MARINA: So, I wrote a book, and in that book I had a chapter on fermenting the revolution.[8] And I use the word fermenting on purpose because I feel it's a bubbling up rather than fomenting: it's non-violent. It's a grass roots revolution.

We need wholesale change. We've got climate change in real time, and we're trying to stay below the 1.5 degree rise in temperature. And in order to do that, we need to change our farming systems really fast. So why do we need to change them? We know that trees and soil draw down carbon. It is the simplest way we know to sequester carbon. There was a white paper from the Rodale Institute in the United States, and lots of other work, which suggests that if all global cropland and pasture were transitioned to a regenerative system, it would sequester enough carbon to soak up the anthropogenic carbon that is being emitted in the world.[9]

Farms also need to decarbonise their processes. Food production and farming currently produce around 30% of carbon emissions. If we transition, we can actually be part of the solution and produce better quality food at the same time.

How can we bring that about? Obviously I'm not a politician, but my simple suggestion is that, as we all eat food, we should all switch to eating regeneratively produced food (local if possible) and eat less meat. We don't have to eat no meat, but if we're going to eat meat, it should be grass fed. If you work in catering – in a hospital or school or a restaurant – you should make the switch to sourcing your ingredients from regenerative producers. If you're a farmer, you should start making the transition.

Everybody can be part of this shift. I've been baffled for the last 20 years why we don't make these changes; I just don't understand why more people don't. Our figures suggest we are sequestering 10 tonnes of carbon per hectare per year here, which means we're carbon negative. Farmers and caterers and people in the food sector need to be bold.

ADRIENNE: Could you talk me through the layout of your farm?

MARINA: Looking at a map of the farm, you can see it's designed so the things that require the most work are clustered around the training centre (labelled Venue), and then it moves out from there with things that need less attention on the edges. Closest, we've got our intensive beds and our polytunnels. Next out are our agroforestry rows along the contours of the field, in between which we're growing our field-scale vegetables. We also have a mobile chicken shed that moves around the farm. To the left, in

The Apricot Centre (illustration: Dilly Williams)

Welcome to the Apricot Centre

Huxhams Cross Field, we have our arable crops, and at the top we have just taken the tenancy on five fields, 25 acres (10 ha), which is in transition at the moment. We'll start farming in there next April with 2 years of wheat followed by 2 years of deep-rooting green manures, and then we'll put the chickens or the cows on the green manures to graze them off.

ADRIENNE: With the extra acres, will it mean you can grow wheat every year?

MARINA: Yes, in the new fields, we're going to have more beef cattle doing mob grazing, rotating with the crops. We'll have up to 300 chickens for eggs in there in the long term.

ADRIENNE: The trees in Billany field, those are your agroforestry?

MARINA: Yes, and they go along the contour. The gradient on the land slopes down from the bottom of the illustration to the meadow at the top. When it rains, the tree rows help slow the rain down and help it penetrate into the subsoil. The village at the bottom of the hill was flooding quite regularly before we came here, and that's stopped.

ADRIENNE: Aside from water control, what do the trees do for the landscape and the biodiversity?

MARINA: We've got biomass hazel: we coppice them every 5 years, and we put them through a wood chipper, and then we put that wood chip onto the soil, we compost it, and that's ramial woodchips which is a great source of fertility. The trees also function as windbreaks and help sequester carbon, of course. They are also home to a huge amount of functional biodiversity. Most of our predators live in there, and they come out into our crops and help with the pests.

ADRIENNE: Do you notice a big difference this year with the drought?

MARINA: We did, even here in wet, old Devon. We had about 10 weeks without water. We collect rainwater and we also have a spring, but we did have to use mains water for irrigation for about a month.

ADRIENNE: I noticed your pond doesn't look very full.

MARINA: That's right. It's slowly filling back up again, but we've decided we need to dig another pond. We are still getting the same amount of water, pretty much, over the year. But it's the pattern of the water that's different: really wet spells in the winter then really dry spells in the summer. We had initially worked out we needed to store enough water for about 4 weeks. Now we're looking at 10 weeks storage. It must have been even harder for you in the east.

ADRIENNE: We had at least the same 10 weeks without rainfall and very little rainfall in September as well. I think other parts of the country did get more. This year I think I've learned just how much water the soil needs. I'm on clay and it does retain moisture, but it gets to a point where it becomes so dry that it's then hard for it to reabsorb the water.

MARINA: As your soil recovers with your regenerative practices, it will retain water better. In the picture, the soil on the left is from when we first came to the farm in 2015, and the soil now is on the right. I think you can see quite easily, if the rain fell on the left sample, the water's just going to run off. Whereas with the sample on the right, it allows the water to penetrate, which drives the small water cycle. This is something I've only just learned about. The trees, planted all the way across the farm, are able to draw water up from deep, from the subsoil, and by providing subsoiling, they let the water go back down when it rains – rather than letting it wash off into our drains. That water evaporating and cycling on the farm makes it cooler and moister.

Apricot Centre soils: (left) original soil from 2015 and (right) from 2022 following regeneration.

ADRIENNE: So keeping water in the ground on the farm?

MARINA: Yes. Conventional training has it that you put drains in to get rid of the water. The drains lead to ditches, and those ditches lead to the rivers and out to sea. But if you keep the water on your farm by storing it in the ground through increased organic soil matter, and by breaking up compaction from below, then the whole microclimate becomes better hydrated and more resilient.

ADRIENNE: It's such a simple solution. You're feeding the locals amazing, healthy food and also saving them from flooding.

MARINA: And bringing all the birds and the insects back. We're heroes. And none of it is particularly difficult or expensive to do.

MARINA: We sell at Totnes every Friday morning, and it's a really special market. We're selling our produce, but we also buy in bits and pieces to top it up. We have a strong food policy that we only source food from Europe and only shipped, never flown.

ADRIENNE: I haven't been buying it in yet, but I think that's something I might consider.

MARINA: What it means is people can come to our stall and buy all of their weekly produce in one go.

Apricot Centre stall at Totnes Market.

ADRIENNE: By the end of a day, do you expect a lot of what you've brought to the stall to have gone?

MARINA: Yes, we never sell 100%. You know the theory is to pile high, creating a feeling of abundance. There's always some left so we hit the restaurants. We've got their numbers, and we call up and let them know what we've got left. Any perishable food like the greens, come Monday morning, we give to a local food charity.

ADRIENNE: How much would you expect to make at these markets?

MARINA: Somewhere between £1500 and £1900 a week.

ADRIENNE: Looking back, do you remember how much you were making in your first year at the market?

MARINA: More like £500 or £700. With COVID people really started shopping outdoors because it was safer.

ADRIENNE: I've struggled a bit at weekday markets with footfall. If people are working on a Friday, how do they still come to the market?

MARINA: There is a Saturday market as well, but that one doesn't do as well. I think people make the effort when you have something they value. They come in their lunch break or they come before work, or finish early and go late. There are quite a lot of mums too.

Winding down

In November I did my last few harvests for the Cambridge Food Hub, which had proved to be a very useful outlet for me over the year. Through them, I'm reaching people who otherwise might not buy from me because they're buying quite small amounts. Because I'm not certified organic, there are some of their customers I can't access yet. The Soil Association certification was progressing; I'd been inspected and got approval to join. From January 2023, I am able to write on my packaging that I was in conversion.

A second attempt at a cover crop went in in late September 2022. The one from earlier in the year didn't take because it was so dry. By November it had come on well and will build fertility and protect the soil over the winter. It will be really interesting to see the difference between covered soil and the land that I have grown on this year when we expand the growing area in 2023.

Eventually all the land will be growing vegetables but with blocks that are set aside. I'll be able to have a rotation where some areas are growing food and other areas are growing cover crops, resting the land between each use, which also helps prevent systemic diseases from building up in the soil.

I was somewhat relieved to reach the end of the year. I was ready for a rest over the winter. It had been a bit of a struggle at times. Because of the heat wave in the summer, a lot of the autumn crops really struggled to get going, and it's actually only now that it's starting to look more vibrant again, which is good because I felt quite disillusioned at times. At least I ended the year on a bit of a of a high with some quality crops coming through. There's a lot that could be improved to make life easier – things to think about over the winter in terms of how to share the workload and to find efficiencies.

It was good to tidy up, to mulch the beds with straw and wool, to tuck them in for the winter.

Looking back

In January 2022 there was nothing growing at the farm. The land was just covered in plastic sheeting and the polytunnel was up. I didn't have a driving licence, so I was getting the bus to the farm. I would cycle to the garden centre with my backpack and fill it with compost and pots. A lot has changed in a year. The big thing was establishing a name. People now know about Sweetpea.

So much of being a small independent food producer is about finding community and attracting support from your community. One of the things I've learned is that if I just talk to people, then the ideas come, the people come, and the support comes. Now I have a year's data, I have more confidence to be able to say to customers when I will have a product and in what quantity.

ONE OF THE THINGS I'VE LEARNED IS THAT IF I JUST TALK TO PEOPLE, THEN THE IDEAS COME, THE PEOPLE COME, AND THE SUPPORT COMES.

We have become increasingly disconnected from the land, and the movement towards rewilding is a bit scary because it sees us as 'other' and separate from nature. Nature is a human construct. We decide what we define as nature. What I have come to experience is that I am a part of the land, I am in relationship with everything that lives and grows on and in the land. Living removed from that connection and the joy that can come from being able to see, grow, eat and share is really hard.

Many people I know who have small holdings around here run other businesses alongside, like a camping business. They need sidelines to keep going, especially because the income is so variable throughout the year. There's almost no money to be made in the winter because not much is growing. Diversification can be really tricky as a small or one-person operation.

There were also a lot of challenges in my journey of acquiring the skills to do this. There are very few opportunities at agricultural training colleges to learn non-industrial farming skills. There are people who are doing amazing work trying to establish training courses, but there's very little support for them. I ended up basically volunteering for 3 or 4 years before I could get a paid job, and that wasn't even growing: it was packing veg boxes. And I had to move around a lot as well. I had to go to a different part of the country in order to get experience because there weren't those opportunities locally.

I'm still learning because last year was so overwhelming with the sheer number of things that had to happen and doing a lot of it on my own, which I wouldn't recommend. I had to not only grow the produce but also set up all of the infrastructure, prepare the land, find markets, find packaging, make a website, go to the markets, all of those things. There's a lot to do. It's a big undertaking even without the pressure to make money almost instantly because you need to cover the bills.

I knew that my family would end up helping, but I had no idea of the extent to which they would. They were there most weekends. My parents were sowing seeds, my dad was putting up the polytunnel, and my sister actually ended up giving notice on her job and coming and helping out for 6 months, which was absolutely incredible. And I don't know what Sweetpea would look like without that.

In hindsight, choosing to do the markets was maybe a bit naive. I think I loved the idea of the market garden and being able to have that relationship with customers directly at markets. But it's a connection that has to be created because it no longer exists, or where it does exist, it's in very small pockets. There are a lot of risks and uncertainties in turning up to a market with a stall's worth of vegetables – the number of hours that go in to doing so are huge. An average market for us would take between £200 and £300, and with two of us putting in maybe 30 hours harvesting, travelling and being on a stall, that results in £5 an hour each, without even considering the costs of actually producing that food – obviously it's not financially sustainable. Marina, down in Devon, was taking twice that at the outset and now six times that amount.

It's very hard for new growers to make money starting out. I think that maybe those who go down the CSA route will have a better idea of their costs and possible returns going into it. They can therefore plan better. But they also often rely heavily on volunteers to actually do the work, especially early on in the project.

It's really hard for me to say that, and I'm really hesitant to put it out there because I want more people doing it. I want more people growing food, and I want it to be a thing that supports people and provides an income they can live on. There are people who are pushing for a universal basic income for farmers as part of the wider basic-income movement.[10] We have an amazing history of allotments in this country, and there is a strong and increasing community of people who grow their own veg. But still there's a strange thing about the value of vegetables – people will happily spend £4.50 on a scotch egg or £7 on a coffee and a cake, but they are horrified if their bag of salad is £2.

But it's not universal. There are lots of people who do see the value of food. You just have to find them, but it is a very strange world to be in because I work in cafes to sustain myself, especially over the winter, and it can feel quite jarring to see how differently we value food.

The biggest challenge I've faced, I feel, is myself. I think I get in my way a lot. I have a perfectionist streak that means if anything has gone wrong, it has been a big knock to my confidence. At the end of 2022 I was incredibly burnt out. I am quite seasonally affected. I find the winter

PEOPLE WILL HAPPILY SPEND £7 ON A COFFEE
AND A CAKE, BUT THEY ARE HORRIFIED IF THEIR
BAG OF SALAD IS £2.

is very hard. I've basically just decided that I am a plant and I need to treat myself that way. I just stop growing in the winter, and I can feel the energy coming back as we come into spring. The things that make me grow are the sunlight, the water and nutrients; I am fed by my relationships with people and community and most importantly with the plants and animals around me and the soil.

> **TOM:** She's been incredible. I've heard visitors ask her how many years she's been doing it for, and then she turns around and says, 'Oh, this is my first year.' And they're blown away. It's almost embarrassing to see how much produce she can make on that plot of land versus what I could have done!

The fact that the income is so low meant that there was a lot of financial stress towards the end. I made the decision not to pay myself in the first year, so I would be in a position at the start of 2023 to actually invest that money back into the business. I now find I'm actually in quite a good position going into year 2. I've been working for a friend's cafe for the winter which has been really good for me because I've managed, through that, to create more links and come up with more ideas. I've been using the cafe as a collection point for my produce, and for 2023 I'm hoping to launch a veg box scheme which people will collect from the cafe and the farm.

Looking forward

One of the things I'm most excited about with starting Sweetpea is the way I'm making connections with organisations and other farmers and producers. I'm stacking enterprise and building communities of reciprocity. A large part of my income last year came through the Cambridge Food Hub, who deliver to local farm shops and delis. They will be restarting in the next few weeks, and they're also connected to the Cambridge Organic Food Company's veg box scheme. So, once I've got my organic certification, I'll be able to sell to them, which will drastically change the viability of my business because they do something like 1,000 boxes a week.

I'm making connections with lots of different people, such as Jacob, the shepherd who grazed his sheep on the cover crops at the farm, who then gave me all the wool from his sheep last year. There's also a local farmer who I buy compost from.

A big change, and picking up on what Marina does, I'll be buying in from the Food Hub as a way of making the market stall more profitable. I'll buy potatoes, carrots, onions – things that I can't grow viably and can't store because I don't have space. This will increase my offerings at the markets and my takings. Also, it's so much less work than growing it yourself. I had my first market last week (March 2023) and that went down really well. I've started selling Hodmedod's quinoa, chickpeas and lentils because I feel that's a really amazing opportunity to support British growers who are producing these whole grains and pulses that add to a well-balanced and sustaining diet. It also improves my offerings and resilience as a business. I've got a friend who grows mushrooms, so I might start working with them to sell those too, and in the longer term, I'd love to sell the flour coming from Tom's grain.

Working with Tom, there's a long-term idea to have more outreach and education at the farm. To bring more people in to visit and share more with our local community. The market I did last weekend was in the neighbouring village, and that's just what I want. I want to be in my local community. There was a really big turnout even though it was raining, and I still hit the higher end of my average market takings. There are opportunities there. I'm hopeful that this local network can develop to the point that Sweetpea can sustain more than just myself. I've a lot of hope for this new growing year. There'll be new life this spring.

Looking back at my finances from 2022, I put in £12,000 of my own money, with which I was able to buy the initial infrastructure – the polytunnel was £2,800; irrigation was £1,050; and all of the other tools and initial materials added up to £7,315. I took £13,260 in revenue – approximately £9,000 at the markets, £3,000 through the Cambridge Food Hub and £680 from wholesale to other businesses.

I'VE A LOT OF HOPE FOR THIS NEW GROWING YEAR.
THERE'LL BE NEW LIFE THIS SPRING.

My expenses were market fees £600, organic certification £400, insurance £320, £1,120 on seeds, £430 on potting compost and £700 on packaging and so on totalling £8,670. From the £13,260 that I took, I made £4,590 profit at the end of the year, which in my first year as a horticulture business, I think is actually quite good. Obviously, it is a very small amount. It's nowhere close to paying myself minimum wage for a year, which would be around £20,000, but I paid no land rent this first year and also lived with family cost free.

I don't see how new entrants going into this without family support or similar would be able to do it. My initial capital investment of £12,000 came from an inheritance, others won't have that. The people who are doing this are making big life sacrifices in order to get these things off the ground.

I think my first-year figures are quite representative of first-time market gardeners. When I was looking for a starter CSA course, one provider shared figures for their CSA start-up (from 10 years back) that showed that in their first year they'd taken £7,000. They've managed to grow that up to a revenue of £90,000, but they had an initial investment of £30,000 to build on. Speaking to other growers on online forums, I think it's actually quite unusual for growers to pay themselves a living wage. Government support under the BPS only kicked in if you were over 5 ha. So, nothing for the small grower. It was reported that the 5 ha cut off would be lifted for 2024.[11] There are some other funding sources, but often grants are

only available to community interest companies or social enterprises and not to sole traders, so I might have to look at my ownership structure. I would prefer not to be a sole trader, but it was the simplest way to start a business, and I just wanted to start.

Other useful data I gathered in 2022 was on my bestselling crops. I was told to focus on high-value crops, especially on a smaller scale. My best performer? Cherry tomatoes. I took over £1,300 from tomatoes grown in just three beds. Then comes leafy salad, microgreens and cucumbers. I'll be focusing this year on the top earning crops so that I can support the growing of everything else.

So much of the conversation around climate change has actually been shaped by fossil-fuel companies. The popularisation of the 'personal carbon footprint' was driven by a BP campaign. It was basically a way to put the blame on us as individuals for this massive catastrophic problem which is not caused by individuals. It's caused by a system that is solely driven by profit at the expense of everything else. If we're made to feel guilty about every single plastic straw and to focus on the damage we singularly are doing, then a lot of our attention is taken up by that. Which can be very demotivating and disabling and leads people to a kind of paralysis in the face of this big, scary thing. We're in the sixth mass extinction event.[12] I'm very scared, but the only way that I can think to move forward is to remember the things that bring joy in life, which are nurturing positive relationships with the planet and working towards those, rather than focusing on the things that are bad.

I've wanted to do this for so long and there's been so many hurdles to jump over. Everywhere that I visit is a reinforcement that I do know what I'm doing. When talking to people who also believe in what I believe in, I hear them supporting me. I think there's a growing movement. The more people see it, the more they realise that it's something that works.

Stacking enterprises, sharing land and working together, rather than competing, is the agroecological approach to farming. We can each take our area of expertise and share customers and share enthusiasm and bring everyone in – rather than shutting people out.

The people who've been doing this work so far have been doing it mostly unacknowledged with very little support. But they have the answers and they have them ready. We just need to support them.

Interlude IV

Food security and UK self-sufficiency

VICKI HIRD

The ability of a nation, region or community to feed itself well is one of the basic tenets of good governance. Yet from the 19th-century repeal of the UK Corn Laws to the global impact of Russia's 2022 invasion of Ukraine, governments have been tested and often found wanting.

A global definition of food security is that it occurs when 'all people, all of the time, have access to sufficient, safe, affordable and nutritious food for a healthy diet'.[1] Food bank growth in the UK shows we've got something deeply wrong. Incomes, wages and welfare are not enough to buy what is, and has been for a considerable time, some of the cheapest food in Europe. Alongside this we have a crisis of obesity and diet-related diseases. The UK imports over 40% of its food (82% of fruit and 45% of vegetables) and climate change is already affecting supplies.[2]

Our self-sufficiency in food has been declining for the past 30 years. Yet there is a strong sense, when 'food security' is discussed, that it means ensuring the current food demand is met through increasing self-sufficient domestic production and by securing imports. But is that somewhat missing the point?

A major political issue

Food price inflation seems to be stubbornly high. The last time global food prices reached this level was 15 years ago, when global instability and food riots contributed to government collapses in north Africa and many governments started to protect food supplies. The peak in prices was caused by multiple factors including oil and energy price hikes, the use of crops for fuel not food, dietary shifts towards more meat, and risky food-commodity speculation. All these factors remain and have been worsened by climate instability, pandemics and wars.

Food prices and supply security are receiving enhanced political attention, and this often leads to greater calls for food security or self-sufficiency targets, and arguments that security should override other concerns, such as nature protection and climate change. The UK Agriculture Act 2020 did mandate the government to produce a Food Security Report every 3 years, and the 2021 report was not bad in detail and scope, though it failed to cover nutrition well.[3] Some are suggesting it should report annually.

False solutions and waste

Systemic issues need systemic solutions, and not just at times of crisis. Deep flaws are embedded in our current food system, which require joined up, cross department, international analysis and action. Siloed and short-term thinking, for instance, with a focus on calories or a single climate metric, will deliver poor solutions. A good example of narrow thinking would be trade agreements that allowed more, but inferior, produce into the country (see Interlude III, Subsidies, Brexit and trade). Or prioritising cheap chicken protein in place of beef to reduce the methane emissions associated with ruminant production. While we do need to reduce methane emissions and fast, such a move would neglect the catastrophic use of land, threats to nature and animal welfare and threats from antibiotics that are associated with intensive poultry systems and the ghosted acres growing their feed.

Another mistake would be to believe global food agribusinesses are the key to nutritional security. The past 50 years have seen major consolidation in food trading and production. A handful of companies now control most global genetic and seed resources, farm inputs, grain trade, food processing and farm tech. This means consolidation of power, both financial and political, is in the hands of a small number of corporations – resulting in huge health, environmental and social problems. It has also meant trade and even aid policies favour monoculture, high-input, export-led production. Vulnerability is built into such systems. The uniform monoculture approach is at far more risk from pest, disease and climatic extremes and is dependent on chemicals and confining vast numbers of overbred livestock, resulting in a higher need for medication.

Food waste also remains stubbornly high despite decades of attention. It seems incredible that over 30% of the food produced is wasted – an unacceptable squandering of land, energy, nutrients and sentient livestock. Waste is now built into the 'just in time', standardised, complex global supply chains that dominate our supermarket shelves. Mandatory rules are urgently needed to put an end to industry practices of poor ordering, the unnecessary cosmetic rejection of perfectly good food, as well as wider food waste by society as a whole.

The international panel of food experts, IPBES-Food, concluded in a recent report that 'agribusiness as usual' and productivity at all costs is not what the world needs.[4] We need diversity more than ever, with climatic weather extremes kicking in ever faster. We should not be using land for wasteful products like intensive livestock feeds, biofuels and junk foods.

Nutritional versus food security

In these modern times there is an acute need to consider nutritional, rather than merely food security, given the crises – both health and economic – well associated with the unhealthy foods that now dominate our diets. Over half (54%) of the calories consumed by UK adults come from ultra-processed foods.[5]

Achieving food security could include continuing an unsustainable supply of foods that are high in fats, sugars, and oils and feeds for industrial meat fed from precious cropland. Nutritional security, on the other hand, considers the health value of food and the ways in which the food system determines an individual's ability to obtain essential nutrients and not just calories. This would mean challenging the huge profits companies gain from selling highly processed, low-nutrition food, and recognising what we need to live well and healthily – not what marketing and advertising lead us to 'want'.

Delivering nutritional security

To achieve the right kind of nutritional, sustainable food security, we need a systemic approach that will follow some key principles.

- Strong and well-enforced protection of vital natural resources. Protection should be based on agreed targets, regulation and fiscal measures, such as taxes on harmful practices.
- Support (including incentives and penalties, advice and research) for all farmers to transition to nature-based and fairer agroecological farming systems – to ensure sufficient, diverse products, import substitution and good livelihoods, while restoring and maintaining the natural systems in and around the farm and reducing climate impact. These farming systems can feed us if we shift diets and stop poor crop use and waste.[6]
- A well-managed, national dietary shift to the consumption of less and better meat and less processed food in affluent countries like the UK – given the unsustainable levels of pollution and land use, and public health risks (communicable as well as non-communicable) involved in industrial livestock systems and high calorie, low nutrient foods. Better meat and food are possible.[7]
- Addressing the true drivers of poor access to healthy, affordable food –

such as rising energy and housing costs and poor wages and precarity – by improving employment standards and fiscal and social welfare policies, so people can afford and access decent food.

- Building up urban and peri-urban food-growing enterprises to deliver agroecologically grown fruit and vegetables to nearby markets. More peri-urban market gardens[8] around towns and cities could also provide new enterprises, training, jobs and green urban places, alongside natural assets such as water management and urban cooling.

- Rules to reverse the huge power imbalances endemic in much of our food supply chains, to ensure farmers can transition to agroecology. The over-dominant, centralised, bullying and multiple retail chains are no longer fit for purpose and farmers need better, more direct routes to market – and infrastructure to match.[9] Public and private investment should relocalise food systems, so they can better link farmers and consumers, remove extractive intermediaries, reduce waste and allow affordable prices and decent returns for farmers.

- Integrated land-use strategies should drive better land-use prioritisation. This should include ending wasteful ecofuel production. Wind and solar production generate far more energy per hectare and decarbonisation should be the priority.

- Curbing the marketing power of food corporations selling highly processed, unhealthy foods that rely on cheap raw materials. The UK has made a start with supply chain binding codes of practice, sugar taxes and curbs on advertising to children, but we need more action.

- Tackling the impact of commodity speculation regarding food will be hard, but it is vital to reduce instability and food price hikes that result from deregulated, self-serving financial markets.

- Putting control of the means of production, for example, seeds, genetics, inputs and land, back into common ownership and farmers' hands is also vital.

- Creating transnational solidarity on food issues: providing financial and other support to enable countries across the world to move away from global junk food production and instead engage in resilient agroecological farming practices. This should include investing in strong early warning systems to avoid worst-case scenarios, and providing support for lower-income countries.

These are major systemic changes and will require significant political will. The ideology of unfettered free trade, and the apparent abundance seen in our supermarkets, are powerful counterforces. However, for many the reality is ongoing malnourishment, and we are on borrowed time.

6 Pasture-fed beef

Treveddoe, Bodmin, Cornwall

BEN THOMAS

WE'RE ON THIS PLANET FOR A PRETTY SHORT
AMOUNT OF TIME AND, AT THE END OF THE DAY,
IF YOU CAN HAVE A POSITIVE IMPACT, THAT'S HUGE,
HUGELY IMPORTANT.

IF I CAN IMPROVE THE SMALL AREA OF THE
COUNTRYSIDE THAT I MANAGE, THAT'S A REALLY
NOBLE THING TO BE ABLE TO SAY.

FARM PROFILE

YOUR NAME(S)
Ben Thomas

BUSINESS/FARM NAME
Treveddoe Farm

REGION
Cornwall

BUSINESS TYPE
(e.g. mixed farm, smallholding, forestry)
Cattle farm

SIZE *(land area owned/managed)*
75 acres (30 ha)

PRODUCTS *(principle business outputs)*
Beef

RAINFALL *(average annual)*
1610 mm

ALTITUDE
210 m

SOIL
Medium loam

**YEAR ESTABLISHED AND/OR
TOOK CONTROL/BOUGHT**
2022

ANNUAL AVERAGE TURNOVER
£30,000

EMPLOYEES *(include yourselves)*
2

LEGAL FORM OF OWNERSHIP
(company, sole trader, trust etc.)
Sole trader

TENURE
(of land ownership)
Tenanted (share farming)

YOUR APPROACH
(self-defined, as you see it)
My approach is to run as simple and as low cost a system as possible. Growing and finishing beef cattle, on diet that consists of grass and hay, which is grown on our farm. Cattle are kept outside year-round, so they gather their own feed and spread their own manure. Contractor costs are kept to a minimum. Inputs are kept to zero. The only thing coming on or leaving the farm is the cattle. Time is in short supply so there is one group of cattle that are moved daily around the land. Simple.

**KEY AGROECOLOGICAL
PRACTICES**
Mob/tall-grass grazing pastures
Woodland grazing
No fertilisers or sprays, farming to organic principles

WEBSITE
https://treveddoe.co.uk

We farm at Treveddoe, on Bodmin Moor in Cornwall. We've got 75 acres (30 ha) that we rent and manage, and then we also graze 55 acres (22 ha) for a neighbour. It's all permanent pasture, all very hilly ground, quite steep, not suitable for growing crops. We have a big chunk of woodland as well. We graze over the pasture with a small group of native breed, Belted Galloway cattle, starting with 20 and building up to 50 over the course of 2022. Cornwall is quite hilly, especially in the north of Cornwall where we are, lots of pasture. We get a lot of rainfall, well over 1 m of rainfall a year. The grass grows through the winter, so cows can stay grazing all year round, and sheep as well. It's a good, really productive area, a little bit milder than other parts of England.

This obsession with cows

My earliest memory of farming with my dad would be in the summertime – the harvest. Making the silage or the haylage, we'd be riding around on the tractor with him, and then Mum would have to put us to bed. And I remember jumping up with my brother, looking out the window and seeing Dad working out in the fields, into the dark, into the night. That really captured my enthusiasm for farming.

I grew up on the family farm. For a while Mum and Dad had a rented farm, and Dad worked on other farms on weekends. We used to go there and help out and, you know, get stuck in. In 2002 they bought their own farm in Wadebridge, north Cornwall.

I have a passion for cattle, that grew and grew as I got older. I think I've become quite obsessive about cattle and grassland. I want to be the best I can at farming, managing and breeding cattle. That obsession grew as I started learning. Seeing how you could do things in a better way or how you could breed better cattle. When I left school, I just went straight into it and didn't look at anything else. I just wanted to work with cattle. It's the cows' nature, how they interact with me. You get that good relationship with them. A lot of cattle have quite individual characters. You know the ones that will always come over for a scratch. They're quite hardy animals as well. They look after themselves. I like that. They can sort themselves out. They don't need loads of shelter, they don't need to come into a shed, which suits how I want to live my life because I want to farm, but I don't want to do it 24/7.

As you go through the seasons and through the years, the calves turn into your breeding stock. The progression of breeding is a really good motivator because you can see how your decisions are affecting the balance of the herd and how it performs. A lifelong goal for me is to have a top breeding herd of cattle. I'm doing small things every year to build towards that, but that's my life's ambition, to have this herd of cattle at the end of my life that I've bred.

After university I worked in Shropshire on a dairy farm which was extensive. We calved the cows in autumn, and they grazed outside for the majority of the year, only coming inside in the winter. That was where I really grew a passion for grassland management. I could see that, when the cows went into a fresh field of lush grass, the milk production went up that day. So you could see in real time the benefit that good pasture management was having on production.

After that, I worked for Beeswax Dyson, a large beef, sheep and arable mixed estate in south Gloucestershire. They used a lot of technology to improve performance and production on the farm. It was really intensive. We were using a lot of artificial fertiliser to boost grassland production and cereal production. The farm was all about modern technology and using inputs to boost production, and I could see that it was a lot of work and we were heavily involved. It always seemed busy. It made me step back from really intensive agriculture as a career path because we just seemed to use lots of time and inputs producing mass food.

I then had a very different opportunity down in Cornwall. That's when, in 2019, we moved down here to manage a conservation-grazing herd for Natural England on Goss Moor, where I could expand on my grazing experience and take that next step in my grazing-management journey. We run a regenerative model, using cattle as our main tool. The cows stay outside all year around; they never come inside. We manage the land so that there's enough there for them through the winter, so we don't have to feed them hay and use straw to keep them inside sheds.

Two jobs and a 50:50 split

In 2022 we chose to come to Treveddoe, which Claudia, my wife, and I run in our spare time. I mainly manage the farm at Goss Moor. It's primarily managing the cattle, a breeding herd of Belted Galloways that graze over the 600 hectares of Goss Moor National Nature Reserve. That's my 9 to 5 role as such, and that's what earns me my salary. Here at Treveddoe I am the farmer and I'm in charge of it all. I make the decisions. We run a herd of both Belted Galloway and Devon cattle, and we fatten them on a mob-grazing system practising regenerative principles. Initially we put as much capital in as we could and were probably only half stocked. We slowly built things up as we earned extra money to bring in more cows.

I have to do two jobs because Treveddoe doesn't pay; we've only just taken it on so it's early days. We're still setting up our route to market and building our customer base, so we're not earning anywhere near enough money to sustain us. It's such a shame that in the current food system we can be grazing over 100 acres (41 ha) of land and not be able to make a living from it. If we fatten 60 cattle on 100 acres, and we take £500 a head, that's only £30,000. And there are other costs to come out before we're left with any profit. It's really frustrating and sad. But hopefully we're on the cusp of things changing, with the way that we're farming here.

We're very fortunate at Treveddoe that the landowners really want to make it work for us long term. But in the short term, obviously, they understand that we've got these full-time roles. The way we're making that work now is the owners are providing the land, the farm buildings and the house for us to live in, and we provide the labour. We're both putting in 50% of the capital to buy the livestock and 50% of the running costs in the first year to cover our cash flow. Then we've got an agreed way of how to split profit in the future. It's going to be a 50:50 split with the owners and, further down the line, the split will shift more favourably to us. For Treveddoe to become our full-time roles, we're going to probably need to upscale, take on a bit more land, have more cattle, but then also start stacking enterprises. For instance, we're really interested in bringing in some chickens to produce pastured poultry, maybe having a little coop for the cows around the fields, and pastured pigs. Stacking enterprises will bring in diversified income and allow us a broader offering of products to directly supply to the market.

Mob rule

On a lot of conventional farms in the UK, farmers will be feeding grains and concentrates to cattle to boost their growth rates. The cattle are housed inside during the winter to keep the growth rates up. It's a lot more intensive. Farmers are buying inputs, trying to drive that production. They will manage their grass by a system called set stocking, where they leave their animals in a field or set of fields for months, just continually grazing off that short grass. Nothing's growing back, nothing's setting seed. There are no different habitats there. So that will grow a lot less forage than the mob-grazing style. You can grow 50% more grass through mob grazing than set stocking.

Whereas here we're stepping back a bit, reducing our intensiveness by having the right amount of cattle for our land. They graze only the grass that we grow on the farm, and we don't have to buy any cereals in – by just managing the grassland in a certain way. The mob grazing system we're using is all about high-impact grazing for short periods, where the cattle are on a small area of the field for one day, and then we move them on again, and then they don't return to graze that area of a field for another 60 to 90 days. So that means you've always got a high proportion of land being rested. All the plants can recover. They can set seed; they can flower in that time. They can send deeper roots down into the soil – that improves soil structure. The increased photosynthesis from the greater leaf area means more microbes are being fed in the soil. Water infiltration is improved through improved soil structure as well. So, you're seeing less run-off and flooding.

There's habitat for pollinators and insects, and you've got tussocky grass for small mammals, because it's not all grazed down, which then in turn feed birds of prey. For the cattle, when they are constantly moved on, they aren't grazing around the areas that they've been trampling, eating, pooing in for quite a long time. They are constantly eating fresher, cleaner pasture, so growth rates increase with the better forage available. They're healthier because they haven't ingested parasites, which means lower wormer use or no wormers at all. Wormers are very harmful to dung beetles. So we're seeing an explosion in dung beetle numbers on the farm, which is awesome; that's another benefit of mob grazing as they break down and recycle the dung quicker, further breaking the parasite life cycle.

And what's more, as the animals are outside all year around, they're not in a shed where they could suffer from pneumonia or maybe increased chances of becoming lame. We're seeing a huge reduction in antibiotic usage as well.

YOU CAN GROW 50% MORE GRASS THROUGH MOB GRAZING THAN SET STOCKING.

It's a great time to be doing it

Nature-friendly farming is the approach I'm bringing from Goss Moor and trying to practice at Treveddoe. Farming in different ways that are good for the cattle, good for the soil and good for nature as well. I see the cattle doing so well, eating a big diversity in their diet, and I think we need to incorporate this greater diversity diet for them back at the farm. The pasture-management plans that we have for Goss Moor are all about the rare species; and for each area, we list projects or tools that we're going to use to benefit and increase or maintain the habitat for these species. And cattle are always a tool that we use for this work!

It's a really different approach, and I have taken a bit of grief from friends because I'm worrying about butterflies and things like that. But I see this big picture where the conservation-grazing management style we have at Goss Moor is increasing the habitat for lots of species, not just for butterflies, for example. It's beneficial for everything and has a knock-on effect: the more insects we have, the more birds can live here. And that's because there's more for them to feed on. What can you say to your grandchildren? Maybe at the end of your life you can tell them, 'I was part of this project that managed cattle over a huge part of Cornwall and had really positive nature benefits.' That's a really nice thing to be able to say, a really good motivator as well.

Love of cattle and managing cattle brought me into this role, but I've always worked on conventional farms producing food and that was the main aim: producing as much as we can from that area of land. At Goss Moor it's different because the primary focus there is not food production, it's nature conservation. But we've also got food as a by-product of the work, and I can see that they can work together, so at Treveddoe that's where we are making it work together. So that's really pushed me down this regenerative path. I'm seeing opportunities starting to open for me because we're going down this route. It's a great time to be doing it. If you are going to be in farming and learning the trade, going down this career path, why not take this revolutionary approach that's now starting to happen across the UK and across the world. Rather than just doing what we've been doing for the last 30, 40 years, which is evidently not working with the biodiversity crisis that we have in this country and other countries, and with soil health declining everywhere. I say revolutionary, but then I talk to my dad and it's about using a lot of the same techniques my gramps used to use. A lot of these things aren't new, maybe we are farming a lot more like we did in the 1930s to 1960s, more than we've done since the 1980s when there was this huge push for food production. It's an interesting journey that we're going on.

During 2022 and early 2023, I've been tremendously lucky to receive advice from two farmers, Matt Chatfield and Nic Renison, who are taking excitingly different approaches on their land.

In the woods

Matt, who runs a silvopasture system in Halwill, Devon (see case study on pages 214–215), first visited me in May 2022.

BEN: We're really keen on introducing the cattle into our woods. I think there's probably a lot of benefits to be gained from bringing them in. At the moment, the understory is probably dominated by ferns, it's a bit of a monoculture.

MATT: You've got 25 acres (10 ha) to play with here, which is brilliant. I think there's great potential. The wood here is different to mine, which is predominantly hazel and willow, here you have a lot of oak. The basics are the same though, it's all about letting in light, and then letting in ruminants and their poo. And then watching the miracle happen.

With mine it's straightforward because it's so easy to coppice silver birch and willow, they come back quickly. Oak obviously doesn't do that so quickly, I have coppiced some oak and it does come back, but very thick. The way your oak is growing, if you coppice some, let in light, then the others will have more room and be better for it. It's also very important to leave any dead ones, which brings another level of habitat for woodpeckers and insects.

Nature is actually really very forgiving, but you just have to take it gradually to start with. You need to observe and get to know your wood. I watch what the deer do. Do you get deer here?

BEN: Yeah, they pass through.

MATT: I get a bit weird here, bit obsessed with this, but what you find, I think, is the wood tells deer what it wants to be eaten. At certain

Ben in his wood with Matt Chatfield (right).

times of the year, for example, the red deer in our wood have recently had foals, and the willow is springing up and they're eating the willow. Different plants and trees respond well to being eaten at certain times of year.

BEN: The cattle are mad keen for the willow. If they have a choice of grass or willow, they go for the willow.

MATT: So what you need to do is choose trees to take down now. Mark them now, but obviously do it in January or February. Coppice then, as all the juices, the sap, are in the roots. Looking at your wood, I think this was managed in the past. The oaks have been left to grow tall and reach for the light. Which you pointed out might have been for the local mines, supporting timbers and the like.

BEN: We have old access tracks probably to allow them to get in with horses and carts to take timber away, and then there are a couple of old stone walls, which makes me think that they did have livestock in here. Why else would you have stone walls in the wood? They separate the wood into two or three sections. So maybe we can reuse those.

MATT: Your wood looks so planned. There's the track, with oaks either side that look a couple of hundred years old. You've got those beautiful ornamental beeches deliberately left. It's not a wild wood. Obviously, it's now been wild for probably a couple of centuries, but there was management in the past.

There's a chap at the Eden Project, Dan Ryan, who's not so far from me. He came to my wood and just by looking at the plants and trees, the shape of them and positioning, was able to tell me he thinks the wood was farmed between 300 and 400 years ago and then basically left. So, what you say sounds spot on. It's worth getting these experts in, they can tell you so much.

BEN: There are other areas which aren't all oak.

MATT: That's good, so the trees that work really well are silver birch, willow, hazel. Oak will grow back but more slowly.

BEN: In the fields, we are mob grazing and moving them around the farm on the rotation. But now that doesn't happen in the wood.

MATT: But you can, that's what I do.

BEN: So, you cut a piece of the wood, and then bring that into the rotation? And then later cut another piece of the wood?

MATT: Exactly.

BEN: We're trying to improve the biodiversity on the farm, incorporating these cattle and farming them in a regenerative way using the mob-grazing system. This woodland is a large chunk of the farm. If we ignore it, then we're not adding as much as we could do. By incorporating the cattle into the wood and creating a disturbance effect with them breaking up the ground and dropping their dung, it can attract more insects and in turn more birds.

MATT: With my woodland, my uncle owns like an acre and has asked me not to do anything with that, which is great because it acts as my control. So where I haven't done anything, there are six or seven different types of plant on the floor. Where I've been cutting and letting light in, 100+ plants have appeared. So, you talk about biodiversity, that's just happened within 2 or 3 years. I'm quite obsessive about the seedbank.

BEN: So you mean what's already there in the soil?

MATT: Yeah. Just let in the light and have a bit of disturbance, and then see what comes through. Different wildlife want different areas, so you definitely want to keep some dark and shaded, and open up other bits. Rather than fattening my sheep in there all the time, the way I use my wood is like a health spa. Sheep, like cows, if they want to heal themselves, they can do it by choosing the plants which will help them. In the wood there are now 200 odd plants for them to pick from. My farming system is to buy old sheep and fatten them. So, some of them aren't great, health wise, when they arrive. I put them in the wood for 4 hours a day for about a week. I'm never going to get my sheep fattened in the wood, but I'm going to get them very healthy there, and then I don't use so much medication.

BEN: Self-medicating, that's a really interesting point.

MATT: So in this area we've moved on to, you can see the previous management. There are bigger trees on the left-hand side, and then this corridor of silver birch on the right which looks like it was cut about 20 years ago. You've probably got 100 trees here. And it's crying out to be coppiced again, all of it. I would coppice all the silver birch here. And if you wanted to, you could protect them, and within a year they're going to shoot up.

When I first did it, I thought I'd killed the trees. And then my first spring, they started growing. I was so happy. I put loads of photos on Instagram, came out the next day, the deer had eaten it all. I was heartbroken, and then I was like, 'I'm going to get someone to cull my deer!' But then the next year, amazingly, 100 times more growth happened. So, basically, the deer produced 100 times more food.

So in here the birch is a perfect trial size for you to get confidence on before you do anything with the oak. Keep the big standard trees, cut everything else down, within a very short space of time, if you graze it right, the biodiversity will explode.

'If you were to cut the larger tree in the background in January or February it will shoot and end up looking like the one in the foreground.'

Dry July

In 2022 July was one of the driest on record for many parts of the UK. Down here in Cornwall we had a little bit more moisture, not very much. The previous spring, in May of 2021, I overseeded the farm with herbal leys. We grazed it down really tight and went in with an overseeder and scratched really nice diverse herbal leys into the pasture that was there before. The cows love that diversity, they love the different bites that they can get.

Many of the varieties in the herbal leys are deep rooting, and this is really key. It's very beneficial with the really dry July that we had. The grasses burnt up and died off, they couldn't hack the lack of rain.

The deeper rooting herbal leys are so resilient. We've now got chicory, we've got red and white clovers, we've got plantain as well. The herbal leys were the only things that were growing. The deeper rooting herbal leys make our pastures so much more resilient. They can withstand these dry periods and still provide our cows with plenty of feed when the modern grasses are burning up. My plan is also to introduce more native grass

Herbal ley mix

We have a mixture of legumes (clovers + birdsfoot trefoil), herbs (chicory, plantain, burnet and yarrow) and grasses.

Clockwise from top left: plantain, chicory, white clover, red clover.

179

varieties, like Timothy and Cocksfoot, the latter of which is really deep rooting, making it even more resilient. This enables us to keep going, to ride out dry spells. Also the white clover was still flowering in the drought, which is great for pollinators that still have plenty of habitat and feed available in the pasture.

A key indicator of overgrazing is to look for which clover is flowering. If you're giving your pasture a long enough rest period, then you'll see your red clover. If you haven't got much red clover flowering in July, it means you are regrazing it too quickly.

The ability of these herbal leys to cope with drought makes us a lot more resilient, especially as we go forward with climate change and the changing weather patterns we're getting with these drier summers. They are such a key part of our farm now, that every acre of land that I can get herbal leys onto I will.

The deep-rooting characteristics of the herbal leys are so much better than modern varieties of ryegrass, which only root in the top 25 mm (1 in) or so. Chicory will root 150 mm (6 in) or more deep down into the soil. Those roots help to break up any compaction. I look for a chicory plant when driving an electric-fencing stake into the ground. Drive the stake in next to it. It always sinks straight in. The root is pumping carbon and sugars deep down into the soil and feeding the microbes across a larger area of the soil. And then the roots are also bringing back up nutrients, minerals and water, from 150 mm or more down, into the leaves, for the cows to eat.

Chicory root.

If you're a cow which one would you want to eat?

We need to be farming in a way that can cope with these changing weather extremes, and these are a great help and a great tool to do this.

Happy cattle, better beef

The data doesn't lie, our animals in the summer are gaining 1.5 kilos a day. People would be targeting that by feeding their cattle concentrates and grains. We're getting that from a 100% pasture diet. We're starting to prove that you don't need to feed cereals and concentrates to hit these weight targets. We can do it on what's growing beneath our feet for free, not using any fertiliser, using the rain, the soil and the sun.

Something I used to be concerned about was the flies on their faces, and I might have treated them with products like Spot On, which is known to be really bad for bees. Now I see less flies bothering them because of

WE CAN DO IT ON WHAT'S GROWING BENEATH OUR FEET FOR FREE, NOT USING ANY FERTILISER, USING THE RAIN, THE SOIL AND THE SUN.

our pasture management. As they go through our long grass, eating what's down underneath it constantly brushes the flies off their faces, and when you see them grazing, the ones that are grazing haven't got anywhere near as many flies on their faces. So, this is another positive about our tall-grass grazing.

Another key thing is that when we're walking through the cattle, they are calm and relaxed. No one's running away from us. They're super chilled. They're not threatened by us being in here. What that will translate into at the end of the day is a better eating quality of beef because they're not stressed. When I handle them, they're not full of adrenalin, which would then ruin the quality of the beef. We should work with them quietly and be kind and give respect because they deserve that. This morning I had to rush to get to work to cut my hay, but I made sure I came up here and moved the fence as it was going to be hot. The cows then had the bottom stretch of the field under two oak trees and the couple of beeches, just to give them that extra shade today. So they're out of the sun. Some people might just leave it and not do that, but to do these extra little things is important; they'll be better for it and healthier.

The ethos of respect carries right through to the end. So, for instance, when an animal (like the steer pictured) is ready to go, he'll spend his last night in the field. It's a lovely chilled atmosphere there. He's not stressed. It's as calm as can be. This is going to help the quality of the product. It's about having the right amount of respect for him. I don't want him

penned up separately in advance. I'd rather take him in the morning to the abattoir, so he spends his last night with his fellow herd members rather than being alone in a different field – or worse – at the abattoir in an unknown place. It's all familiar ground for him. Nothing's changing. Nothing's different. We'll come up in the morning and pretend that we're moving them into the next block of ground, the next parcel of grass, and instead run them all back to the yard and draft him out, load him up, then bring all the others back out to the next day's worth of pasture – a nice, calm process in the morning.

The local abattoir we use is only a few miles away from where we farm. It's a small family-run place. It's got a good atmosphere. Lovely and chilled, it's the right place to take our cattle to. The steer that we've chosen to go, he's at a good weight now. He's got the right fat condition on him. He's been fattened all summer, 100% out on the herbal ley pasture. He's had no feed, just out on the pasture, and he's fat and ready to go. He's not a youngster. He's over 3 years old, so hopefully that will translate into a really good depth of flavour and maturity in the beef.

We're a small farm, and we're trying to use local businesses and local people to play our part in the local food chain. That is part of our system. That's why we're using this abattoir just down the road. We've been using them for years now and have a really good relationship with them. They do a really good job. We feel it is a better outcome for animal welfare. When the animals go to a smaller abattoir like this, that's a lot closer to the farm, they haven't got to travel long distances. It's really important that we support small local businesses, to keep them going. It's better for our animal welfare, and it's better for local people, for the local economy. It's a win–win. Then we plan on using the local butcher, James Kittow, he's just down the road as well, another good local business. James does a really good job in butchering the meat.

The other local businesses in the food chain are, in general, very supportive of the farming side of it, how we're using regenerative farming practices. Many, like James, are very passionate about that as well, and they think it's great what we do. These guys are really enjoying the fact

WE USE LOCAL BUSINESSES AND LOCAL PEOPLE TO PLAY OUR PART IN THE LOCAL FOOD CHAIN.

that we're supporting their businesses, and in turn they've told me they need to support the future of the industry to ensure that they have a business in the future as well. They've always been really helpful to me, including when getting set up and started, any questions I have, they're always on the end of the phone. We're producing local cattle which is being processed locally and marketed directly to local people – it's just a win for everybody.

Direct selling is gaining in popularity. I think it's something that's going to happen more. Once you've done it, you realise that it's really nice. It's really special to produce food like this and go through the final process as well as taking it to the abattoir, having it butchered, and supplying it to customers.

That's not to say it isn't hard. It's tricky, especially when you are separating them out, and because you've got it all calm and controlled, a steer will walk up into the trailer by himself, no pushing; it can seem like he's choosing to go to the end destination. That sticks in your throat – he's not going to enjoy another lovely day in the pasture.

I sometimes wonder, 'Why do we have the power to send this animal to its death?' But, you know, you've got to think about the bigger picture. This animal's going to go off and we're going to receive back probably 160 kilos plus of beef. If we have a 200 to 250 gram portion size, that's feeding up to 800 people with one meal, 800 meals from this one animal.

That's the food side. Plus the way it's been produced within the regenerative system we have, with the mob grazing, with no inputs at all, with the biodiversity improvements they provide, these animals have been actively contributing to nature while on the farm. These animals have a really good purpose to their lives, and then they provide food for people. If we're going to eat any beef, it should be this beef. I think it's really important. If we stopped eating meat, then we'd lose these animals that do such important roles in managing not just farm pasture land, but conservation grazing like at Goss Moor as well. We can't afford to lose that. We need these animals to keep on top of these places. That's why I am a livestock farmer and I want to keep being one.

IF WE'RE GOING TO EAT ANY BEEF,
IT SHOULD BE THIS BEEF.

Over 800 meals from one of our Belted Galloways.

Meat extravaganza

In the photo, you can see all the beef that came from one of our Belted Galloways. It's pretty awesome to see how much you get back from one animal: over 200 packs of beef, which is 800 meals of amazing nutrient-dense food. A great range of cuts for different meal occasions and in different price ranges so hopefully accessible to lots of people. Our meat is sold direct to customers both locally and nationwide, some to butchers, hopefully also to restaurants in the future. We're having some exciting conversations about where it will be enjoyed.

We're developing a good courier relationship, so we can distribute it nationwide. But our hope is that a lot of our produce will be distributed locally. That's what we'd really like to happen – for it to have an impact on local people, with customers coming to the farm to buy it and seeing the animals and seeing the work the cattle are doing.

For regenerative or nature-friendly farming to become more mainstream and recognised by the public, we need local people to try it, to buy it regularly. Then they'll be investing in our farm and helping us to do the work we're doing with our animals. Selling and connecting locally really helps the local, social-ecological system.

It's important for us to establish and tell our Treveddoe brand story, which centres on how this beef got to the table. We want to be super transparent with our customers, so they know both where it's come from and how it's been reared, and also the benefits the cattle have had on the land and the benefits they get from the land in terms of improved nutrition. Compare that to people just going to the supermarket, taking a product from the shelf and having no idea how it got there.

Well-managed livestock

In August 2022 I went to visit Nic and Paul (Reno) Renison on their Cumbrian livestock farm (see profile on pages 304–306).

BEN: So how big is your farm here and how do you manage your land?

NIC: The farm is 350 acres (142 ha) and we rent another 25 acres (10 ha), and we manage it in a regenerative way. We have rotational grazing and mixed species of livestock.

BEN: And how do you manage that land?

NIC: When we first came here 10 years ago, we were very conventional in our thinking. By that, I mean we were set stocking, so our animals were in fields for long periods of time. We bought in feed, we bought in fertiliser, and we were just producing a commodity with no control over anything, particularly our costs. Moving into a more regenerative system, you claw back control over those input costs, your inputs are very little.

BEN: What does pasture-fed livestock mean to you?

NIC: To me, pasture-fed livestock means that those animals have not received any inputs other than pasture. So that pasture can be grass, silage, hay, but that's it for their entire life. In comparison, you often see grass-fed steaks in supermarkets, but they could have easily been grass fed and also grain fed. And that's a labelling issue.

BEN: Definitely. I think there is a big difference in the product as well, those cattle could be grass fed at some stage but are often finished on cereals. Pasture fed should really encompass their whole lifetime.

NIC: Yes. Generally, ruminants are meant to graze, and we for some reason over the last 60 years thought it's great to feed them cereals. All the cattle and sheep in the UK could be pasture fed. If we did that it would be a massive step in reducing our reliance on grain and freeing up loads of land currently producing grain for animal feed.

Ben with Nic Renison (right).

BEN: Definitely. Hopefully it'll keep growing and there will be increased awareness of the good standards and good techniques that people like you guys are using.

NIC: Farming conventionally is quite easy. So, you have a problem, you go to the cupboard, you get a bottle of drugs, or you get a bag of feed or a bag of fertiliser. To farm in this way, it's a different way of thinking, it's a different mindset. And that isn't taught at agricultural colleges, it's not the norm. It's not really talked about in the press. So you continue to go against the grain, having to go to different places to learn things and also networking farmer to farmer.

BEN: How does the change to a regenerative approach affect stocking rate?

NIC: Well, stocking rate is a measure of how many animals you've got per hectare. Moving from conventional inputs, when you change from using fertiliser to not using fertiliser, it's a bit like coming off

SO YOU CONTINUE TO GO AGAINST THE GRAIN, HAVING TO GO TO DIFFERENT PLACES TO LEARN THINGS AND ALSO NETWORKING FARMER TO FARMER.

drugs. It is common to have a dip, a 3-year cycle where you see a reduction in output. So we did lower our stocking rates for the first 3 years, but since then we've been climbing back up.

Production wise, after you've gone through that initial phase of getting the soil to start recovering and getting everything into sync with nature again, you then can start to climb again. And I think we could even be more productive.

BEN: In regard to stocking rate, it needs to be contextualised. You guys are on a 100% pasture system with your cattle and sheep. But other people might say, well, my stocking rate is X amount higher, but they've got these ghost acres, they're buying the feed, the soya, produced on massive areas of land elsewhere, and that isn't factored in. And that's using no fertiliser.

NIC: No fertiliser, no bought-in feed for the ruminants. This whole current system since after the war was all about food production and that was amazing for Big Ag. So you get a free calendar and mug with a brand on it every year because someone's given it to you for buying into the system. I was recently on Twitter, and I saw a picture of a load of agricultural students at a well-known agricultural university sponsored by Bayer, wearing jackets with the logo. That is a perfect example of how entwined our system has been with Big Ag.

BEN: They've grown up with it. These are the companies they're used to seeing and dealing with. I think it's the norm.

NIC: When you start to look at it differently, it's totally infuriating, and you want to kind of release these farmers from this grip?

BEN: Yeah, definitely. I agree. I get advice from companies, say, seed companies, and they say you need to use their seeds, but you must do this and you can't do that. Farmers probably need to become a bit more independent in their thinking in that respect and do what they want to do to see if it works. Because a lot of these people are just trying to sell more product.

NIC: Salesmen will turn up on your farm, and they'll say hello and then tell you you're doing a tremendous job here. They'll talk about the people they know you know, and then they build trust with you, and then they'll screw you. They say they really want to help your business, but for them it's just selling more.

BEN: So have you got any examples of when this has happened before?

NIC: We've had fertiliser salesmen try to say you can't possibly grow all the crop you need without fertiliser. But we're speaking now in 2022 when fertiliser prices have rocketed, and many of our neighbours have said they're not going to use much fertiliser this year. They're going to seriously reduce it because they can't afford it. And actually, their grass growth hasn't been affected. It's stayed the same. So I think that is just an example of where farmers have been sold this idea that they've got to have this fertiliser, otherwise the ryegrass won't grow. It's just a cycle that they've continued, and when you're in debt and you've got employees to pay, it's very difficult to take the risk to get off that ladder.

You've just got to go to an agricultural show and see the Big Ag machinery and feed companies. They're all driving nice cars, having holidays, good salaries.

BEN: Yeah. And we've not got any of those things!

It just makes our operations so much more resilient if we're not spending the money and more enjoyable to not be in a tractor all the time applying these inputs. Do you think your business is using less fossil fuels? Do you think your farming system here is helping to combat climate change?

NIC: Hugely. We have diesel for the tractor and petrol for a quad bike, and feed for the pigs and chickens, but we don't buy ruminant feed or fertiliser and have all the deliveries associated with that.

As far as climate change goes, this farm is an oasis, a bubble of greenness where we have very low inputs. Our reliance on fossil fuels is minimal. This green bubble needs to become mainstream. But when you look at our farming neighbours and the rest of the conventional farming industry, their reliance on fossil fuels is huge because they need big machinery, they have lots of inputs. It needs a huge amount of energy to get to keep that kind of farm running.

IF OUR KIDS SAY, 'WHY DIDN'T YOU DO ANYTHING ABOUT IT?' I DON'T JUST WANT TO REPLY, 'WE COULDN'T BE ARSED.'

As we've gone further down this regenerative route, we've become far more climate conscious. And I often think that if our kids turn around to us when they're 25, and everything is so much worse than it is now, and they say, 'Why didn't you do anything about it?' I don't just want to reply, 'We couldn't be arsed.'

BEN: Your farm is a great example of a regenerative approach. It's reliant on itself, pretty much just on the land you've got. Except for the pig and poultry feed, it's nearly a circular system, and that's how it should be. It's crazy that farms have gone so far away from that.

NIC: It's very easy for people to think that farmers have done this to themselves, but they haven't. It's been government policy that's pushed them, heavily subsidised, to do destructive things like rip out hedges and massively overstock sheep. When you step back and look at it, policy has been a major cause, and although policy is trying to make things better, it's still got the potential to do destructive things if not changed in the right way.

BEN: Maybe that's what's great about the whole regenerative farming movement in the UK – it's so farmer led.

NIC: Yeah. I think the government are listening, I just worry that this type of agriculture doesn't suit big business and doesn't suit shareholders.

BEN: So what do you think the government is doing at the moment to support farming in this way? And what do you think they could do to promote that going forward?

NIC: Well, I think the new Environmental Land Management schemes (ELMs) have got the potential, if they're brave enough, to be really good. In agriculture, typically, you've got food production and you've got the environment, and I would like to see ELMs being the tool that brings those two together.

We've been involved in some of the pilots of that, and I just hope that they stick to their guns, and it becomes a living thing. Because I think that you have to look globally, there's the environmental disaster happening, there's a health disaster happening and there's a biodiversity disaster happening. Farming in the conventional way creates a lot of those problems, but farming in a regenerative way solves a lot of those problems.

BEN: What we're doing now is not working.

NIC: I think we've got to think of farming as food and biodiversity and nature; all those things rolled into one.

BEN: One way that government could help is by breaking down the barriers that stop small producers being able to process food on farm. I think you've had an experience of this.

NIC: Yeah. The beef animals travel an hour to the abattoir, and then they come back to our local butcher. She does a great job. But it's a long-winded way of doing it and it's quite expensive. So, wouldn't it be amazing if she was subsidised? That would mean we could sell at a more affordable price, and it would mean that more people could get trained as butchers. It's a real gap currently, there aren't enough small-scale processors and abattoirs. The hurdles to processing our own pastured chickens were just too great.

BEN: It's set up for the big players, isn't it? If you want to process your own beef or some chickens on the farm, you've got big hoops to get through which make it unrealistic.

NIC: I get to the stage where I don't think I've got it left in me to fight against the Environment Agency, who don't really get it, don't really care if someone down the road wants a local chicken that tastes of chicken. They don't really want to make that happen. We have big factories where food is produced and it all goes out from there – it's bonkers.

BEN: What does a beneficial circle mean to you and your farm?

NIC: The beneficial circle is a phrase that is quite new to me. Over the last 10 years we've become much greener. Now we're starting to think of the farm as an ecosystem. So to give you an example of that, we've just planted lots of 5 metre wide hedges, partly for animals to browse on, partly as habitat and partly because in 10 or 15 years they will provide much more. We can cut them, chip them,

WE'VE GOT TO THINK OF FARMING AS FOOD AND BIODIVERSITY AND NATURE; ALL THOSE THINGS ROLLED INTO ONE.

and then put that woodchip in our muck, which we now compost. We've learned a lot about composting over the last few years, we need to up the fungal activity in our compost so it's more balanced with the bacteria.

So, the beneficial circle is using the different enterprises on the farm to talk to each other. Coming from a monoculture dairy farm background, just ryegrass everywhere, I never thought I'd be talking like that. If you were an 85-year-old from a farming background, this would make sense to you because that's how you used to do things.

BEN: With your chickens following the cows around the farm, how does that work?

NIC: Three or four years ago, inspired by an American farmer called Joel Salatin, we built an egg mobile. The idea is that we have a lot of hens in our egg mobile, and they follow the herd around the farm. They follow 3 or 4 days behind the cows. They sleep in their egg mobile at night, and then in the daytime they're out from dawn to dusk, and they scratch all the cow poos and just do a great job. It's a really perfect symbiotic relationship with the cows.

WE HAVE BIG FACTORIES WHERE FOOD IS PRODUCED AND IT ALL GOES OUT FROM THERE – IT'S BONKERS.

Renisons' egg mobiles.

BEN: They have an amazing life.

NIC: And amazing eggs! The hens have soya-free, palm-oil-free, GM-free feed. You let them out and they just start just eating grass, I think 30% of their nutrition comes from grass, and who knew that chickens ate grass?

BEN: And then breaking down those cow pats as well, dispersing them into the ground quicker.

So I guess it's not such a big change for the chickens to go around the pasture.

NIC: Exactly, it doesn't cost a huge amount of money to set up. So as far as getting a new enterprise onto your farm, it's not a massive investment. And if it absolutely goes wrong and you decide that you actually hate chickens, it doesn't really matter in comparison to spending £200,000 on a new poultry shed.

BEN: Shelling out on a mass of metal and concrete which you then need to earn a bigger income to pay for.

NIC: And the eggs are a great 'gateway drug' when you're selling food locally. If people like your eggs, they'll then try your meat. So it's a low-cost starter to get people thinking about buying from you.

BEN: Are there any other benefits of the egg mobile for the cattle?

NIC: The chickens sanitise the pasture, so they go in and they go through the cow pats and eat all the fly larvae. Flies are a problem for cattle, so historically we've used some insecticides, and now we don't.

Wouldn't it be great if every dairy farmer had 300 or 400 chickens following the cattle, and then more of our eggs could come from this way of farming, and less chemicals would be used on the cows. We'd also be feeding the public better quality eggs. The possibilities are endless, and it's not until you've spent time in industrial free range sheds that you realise it's not normal in any shape or form to have thousands, up to thirty thousand chickens, on one pad of concrete.

BEN: It's going to be healthier for the chickens as well and reduce antibiotic usage.

NIC: It's all good. It just requires people to be brave enough. For big milk companies to say, well, part of our story is that we're going to have these egg mobiles following the cows.

BEN: It's a great thing if you can go to the end of your local farm track and get a pint of milk and some eggs – two of the main things people go to a supermarket for. If you could get that locally all the time, then it would be very powerful.

NIC: We've also got a very small pig enterprise, and they just pulse through the woods. They come into areas of the farm, woodland and little corners of scruff that we've got, and they're in there for quite a short time, maybe a week. They turn everything over, and then don't go back into that spot for a whole year. They kind of reinvigorate things and cause chaos.

BEN: How does that benefit the farm and the environment?

NIC: They renew it, refresh it. They have a massive impact, and then you need an enormous rest period. They encourage new life and are moved on before they have wrecked an area. That's why we only have a few pigs. It's why outdoor pig farming is difficult to manage and why most are housed inside.

We take the pigs to between 9 and 12 months, and then they are made into sausages, bacon and chops, and sold locally. I think we give them their best life, and then they taste really good. You get a lot of meat off a pig.

BEN: How much bought-in feed do they need?

NIC: They are on about a kilo a day each.

BEN: If you left them to their own devices, no bought-in feed, would they be OK?

NIC: If they had a big enough area, they'd be fine eating roots, windfall apples and the grass. I think at Knepp they say 50 or 60 acres (22 ha) per sow, it's a lot of land.

BEN: Looking at your set up, all you have, I'm really exciting to see food produced in this way, especially with the chickens and pigs as well. I've got a bit of experience with cows and their rejuvenating ability. Looking ahead I can see how great things could be. With the changes you've made, have you seen a change in the health of your livestock?

NIC: When we first of all came, we wanted to have loads of sheep. We had 1,200 sheep at one time, and when you've got so many of any monocultural thing, you just get battered by illness because those animals are all the same and kept quite tightly together. Infections just go through them, and that's the same when you've got thousands of free range hens. The way we do things now is much more beneficial for animal health. We just use antibiotics in a targeted way, such as a calf with pneumonia recently. We don't use blanket treatment for anything. It's better for animal welfare and financially it's better too. And also it benefits our customers, who are buying beef which has never had antibiotics. Emotionally it's better for us too. It's quite stressful when you've got sick animals.

BEN: A minimal level of intervention as well. So that's great.

NIC: And, you'll be aware, working with animals every day, when they are thriving and they look good, and you feel good, everything is happy. And it's also about having the right level and diversity of stock. We're not there yet, but we're working our way towards achieving that balance where generally everything is thriving and working with wildlife.

BEN: How do you market your produce?

NIC: The eggs, the chickens and the pigs are all marketed direct to consumer, but in relation to the cows, they're small enterprises. With the eggs, there's an egg club. I've got villages that I sell to on a subscription. The beef, we sell some of it in boxes direct and we sell some by post. But the majority, probably 90%, we sell as store cattle to a chap who is Pasture for Life accredited, and he then finishes them on his herbal leys. That's quite a new thing for us. Historically we were selling them into the market.

My message on the selling is that just because you farm regeneratively, it doesn't mean you have to sell everything in a box. You need to think about your margin. You may not be capturing the top price for your produce, but there's lower costs, so the margin can be higher.

BEN: Would you expand your own beef finishing?

NIC: I think we will probably grow it. I would like to have a couple of butchers that stock our beef. But you need a near constant supply for that. Dealing with those lean months in January, February, March, when you would struggle to finish pasture fed, is the problem. But what is amazing is when you do sell products direct and you get the feedback.

BEN: It's a very social, affirming response.

NIC: Looking at the wider agri-food system, the danger with local direct selling is that you only sell to people with more disposable income, to those people who can afford it. And that's where the subsidy needs to come in, to allow us to sell it locally at an affordable price. To have a beef animal processed costs £400–500.

BEN: And then your time to package and market it.

NIC: But you want all local people, those who maybe have not bought local food, to have some produce from a pasture-fed system and then not want to go back to a supermarket. Wouldn't it be amazing if the supermarkets sold more pasture-fed meat?

BEN: They have the power to do it. They could affect really big change.

NIC: The supermarkets hold the key to so many things, probably more than government really. They have such footfall. And they're so powerful.

BEN: Has the farm become more profitable with your transition to a more nature-friendly model?

NIC: When we first got here, we were lucky enough to be able to buy the farm, but part of buying the farm was that we were saddled

THE SUPERMARKETS HOLD THE KEY TO SO MANY THINGS, PROBABLY MORE THAN GOVERNMENT REALLY.

with a huge mortgage, and that mortgage actually has sharpened our minds to think differently. If we'd have arrived and it was all paid for, we would have been quite lazy in our thinking, and I don't think we would be doing what we're doing now.

All the time, I'm thinking about cash flow and paying bills and kind of juggling things. Very early on we wanted to buy loads of machinery, and we just held back from doing that. We've had to be quite tight. So the egg mobiles, for example, they've cost thousands rather than tens of thousands, and we're now, ten years on, in a much better financial place. Our Basic Farm Payment used to be £25,000 a year, now it's £18,000. So that's going down. When we first arrived, we both worked off the farm part time, and then I had a full-time job for a while. So, we've had to support it with other money, but now we're at a stage where everything is much easier. We've just gone into a Countryside Stewardship scheme on one area of land and it's getting easier.

I think with farming, you never do it to become really rich. But what we've done is stack enterprises. For instance, we have a fledgling glamping enterprise. And so it's about layering things up. If one source of revenue is down, another one, hopefully, should be up.

BEN: So are you still working off-farm?

NIC: I'm here all the time now and that's great. For a while it was quite frantic working off-farm as well. Now we are getting to where we need to be, life's a bit easier financially.

BEN: That's really good.

NIC: So financially our priority is to pay our mortgage off, then that will allow us to invest in other areas of the farm.

BEN: If you're living the life you want, then maybe it's not about money. It's about lifestyle and achievements, the things you want to do. If you're doing those things and if you're happy, then you don't need so much money.

NIC: We feel quite lucky. We can go for a walk over our own fields with the dogs, we've been able to bring up our kids in a lovely environment, so as far as benefits go, I think they're huge, but they're not necessarily financial. We're not into consumerism, so I don't feel we're missing out.

BEN: What is Carbon Calling? What's that all about?

NIC: In 2019, four of us (Reno, myself, Liz Genever, who's our partner in Carbon Calling,[1] and Tim Nicholson, who's the host of Carbon Calling) all went to Groundswell and just loved it, but we thought it's quite arable based and it is blooming miles away from Cumbria, so wouldn't it be great to do the same up north with more of a pasture-grazing angle to it. So we started organising and then COVID hit. But this year, in 2022, we finally did it. And it was amazing. We had 200 farmers in a shed at 9 am on a Saturday morning, the place felt alive with positive energy! From the success of that first event, we're now thinking we're going to do it again in September.

Why it works so well, and this would go back to the early Groundswell days, is a relatively small number of people engaging in farmer-to-farmer learning.

BEN: The way you base it on livestock is going to pull in loads of people like myself. Because Groundswell is very arable focused. Livestock production, and agriculture in general, contributes to greenhouse gas emissions. What are you doing here to mitigate this?

NIC: I find the carbon calculation quite confusing and annoying, but I do recognise that livestock can be a problem. How I rationalise it is that we are increasing our biodiversity. We've done loads with hedges. We don't use any fertiliser, that's a massive one, and we don't use any bought-in feed for the cattle. And we've got a grazing regime where we have long rest periods, building grasslands that sequester carbon. I have concerns that the calculators don't record such things very well.

What I know is that when you're with our cattle, there's just loads of life around them. There are swallows and other birds flying around, insects and dung beetles. I see them as part of the whole biodiversity system. If we were farming conventionally, and those cattle got put in a shed and fed soya, we could in fact finish them quicker, and the argument for an intensive system is that, well, they're alive for less time so breathing and belching and trumping less. But I can't get my head around the fact that we destroy the Amazon to produce soya, to get it over here to feed them, and there are none of those wider benefits.

I think part of the issue is agriculture being used as a scapegoat by just looking at cattle. But I do recognise that industrial farming is incredibly bad for the environment.

BEN: It's not looking at just carbon emissions. If you look at just one thing, you're going to ignore the other things that are vitally important such as the biodiversity.

NIC: For some reason, cows are a particularly good thing to bash at the moment. There's that saying, 'It's not the cow, it's the how.' It's very different on a feedlot in America of course. But we're very different to that industrial farming. Part of our ethos here is that anyone, any time, can come and see what we're doing.

BEN: Do you do any monitoring here of your soils, water or biodiversity?

NIC: Yes, we do, now. If you were just starting, I'd say do loads of baseline data, because we didn't and that was a mistake. We now do lots of soil testing, we've had bat surveys and people come to look at insect numbers and flower species counts.

BEN: Big question, why do you farm, Nic?

NIC: I was brought up on a farm, and I think it's a very honest way of making a living. You get up every day, and you look after the environment and make food for people, good food. I've had other jobs where you come back in the evening, and you think, what did I actually achieve today? So, I think job satisfaction is very high and generally, most days, I enjoy it and it doesn't feel like I'm going to work.

And I think the whole environmental challenge, the climate catastrophe, we can help solve that. If you just live in a normal house you can, through your buying decisions, make things shift a bit, but you can't actually change stuff like we potentially can. And that's exciting.

IT'S A VERY HONEST WAY OF MAKING A LIVING. YOU GET UP EVERY DAY, AND YOU LOOK AFTER THE ENVIRONMENT AND MAKE FOOD FOR PEOPLE, GOOD FOOD.

BEN: So, what does soil mean to you?

NIC: Soil means more and more, I never thought I would get excited about soil. I just thought it was a thing that things grow in. When we moved to the farm, we did it without digging one hole and soil wasn't really on our radar at all. By pure fluke we bought a farm with quite good soil. And now we look after it. It's exciting that the better you treat it and the more you take it into consideration, the better it's going to get.

BEN: And what is regenerative agriculture to you?

NIC: For me, regenerative agriculture is working the land so that, when we've finished here, it's in a much, much better condition than it was when we started. Every day you're looking to improve soil health, improve biodiversity, improve nature, improve the quality of the food you produce. It's building on that every single day and doing all that without artificial inputs. It's a continual gradual uplift in all the good stuff.

We're doing Carbon Calling and all the regenerative things we do on the farm as we really believe that this is the way forward. It's a grassroots movement that puts farmers in charge of their own destiny. It benefits farmers and the local community first. And then on the back of that, you've got huge benefits for society and people's health and the whole world.

Back in the woods

In January 2023 Matt Chatfield returned with me to the area of wood he suggested we start our coppicing in.

BEN: A few weeks ago, I was looking into the history of the name Treved-doe, and it means the farm with the birch tree on, which is rather cool. We came here last May, and you picked out this as one of the good areas to coppice. But we waited till now to do it, in January. Why is that?

MATT: Basically, during the winter period, the sap all goes into the roots, that's all the energy. So, if you cut it in the middle of the summer, you would essentially kill the tree because all the sap would be in the trunk. Whereas now the sap is in the roots ready to go in the spring. Cut it now and it'll start springing up lots of new regrowth.

BEN: Why are we choosing these smaller birches and hazels instead of, say, a big oak? To my untrained eye, I think big tree, cut that down and you get a lot of firewood and let in a lot more light. But is there a reason why we're targeting these instead?

MATT: Personally, I think you need to thin out. You've got some lovely standing oak here. You want to leave that and work with the previously coppiced birch and hazel, those last cut about 20 years ago. Let the light in and then get your animals in. That is what I'm doing on my land.

Try it in this area, see if it works and then if it does, roll it out over the rest of the wood in stages. You see there are loads and loads of leaves on the ground. When I came in summer, this was totally covered over, there was no light getting in. Some of these leaves are probably from the winter before. They're still there because there's no light, there's no trampling, there's no actual activity happening in the soil.

BEN: So how do we kick start things?

MATT: There'll be seeds that have been there for many years. All you've got at the moment are ferns and bracken. They're growing because nothing's competing with them, they love low-light conditions and low fertility in the soil. To try to compete with that you need to add fertility and to stir things up. You just want to create nice patches of light. You want to cut low down with an angled cut so the water will run off and—

BEN: —a nice clean cut? I was reading the other day you don't want to create spaces, nooks and crannies, for the bacteria to get a foothold.

MATT: One thing I've noticed with my hazel coppices, in my control area, where I've not cut and there are no sheep, there are no hazelnuts, but where I've cut and put my sheep in, there are hazelnuts on the trees. My thinking is the changes and increased diversity are slowly bringing in pollinators. More light, more nature.

BEN: It's really exciting stuff, isn't it? I think a lot of people for a long time have been shutting off woodland, not putting animals in. I love how there are people like you, and hopefully we can do similar, putting the livestock back in woods and using this good asset. And how good it will be for the cattle to come in here and have a bit of a rest from the weather.

MATT: And it's also very good for farmers. I've got bluebells appearing, I've got primroses, I've got dog's mercury. And after a long day, it's my treat to go into the woods. For mental health it is brilliant.

BEN: Yeah, that's a really good point. This will become a lovely sheltered glade, and it will have its own little microclimate, will stay slightly warmer. So, at the moment the cows are having about half an acre (0.2 ha) of grazing a day and a whole round bale of hay.

MATT: Are you going to put a bale in here as well then?

BEN: No, I'm thinking that I want the cows to have more of an impact in here. So I'll give them access to 1 acre (0.4 ha) of woodland.

MATT: Let's get some stakes in and wire up and get the cutting done.

BEN: Time to get some cows in.

BEN: It feels like it's where they're supposed to be.

MATT: It's amazing to see this. Is it not the most natural thing you've ever seen?

BEN: And what these woods have been missing. Already there is dung and hoof marks, where the cows have been squashing leaves and organic matter into the soil.

MATT: Look at them eating the ivy. They came in, crashed around, scratched and now every single cow is eating.

BEN: When you see cows in fields and all they have is grass, especially modern rye grass, that's all they've got to eat, compare that to this. They've got ivy, ferns, willow. Cows don't want to eat just grass.

MATT: I guarantee that they're getting minerals that they just haven't got elsewhere. I always think with animals that they get themselves into balance. We can give them mineral buckets, licks or whatever, but if it's available they take what they need. And I actually think it helps them to get fat.

BEN: There's another great reason for having livestock – cows, sheep and pigs as well. You can use them to target problems. If this was completely overrun with brambles and bracken, you could send a group of cows into one area to really knock the plants back. Then maybe if you wanted to create some more ground disturbance, that's when you could bring pigs in as well.

MATT: It's blowing me away how much food they're finding at this time of year. It's quite clear that this is a suitable time of year to bring cattle into woods.

BEN: I'm thinking that I need to change the whole-farm management plan. How can we get them in here sooner? And what help can I access to make this a part of my resilient grazing? It is a huge tool to respond to climate change because when it's absolutely horrific weather, like lots of rain or intense heat, I can bring the cows in here.

Three little pigs

Now we also have our three Tamworth cross Oxford Sandy and Black pigs. I brought them to Treveddoe after we visited the Renisons' farm in Cumbria. I was so impressed with how their pigs work in the woodland across their farm. I thought, 'Why are we waiting?'

The pigs have come through and cleared all the acorns up and really turned this place over. The understory of the woodland was just dead, no diversity in the sort of plants that were growing here. I want a lot more diversity in here, and these guys are the key to that. Their action of walking through and rooting and trampling is going to start bringing the soil back to life.

We've got this good overstory of oaks and other established trees. But I really want what I saw at the Renisons' farm in Cumbria, which was amazing. They had waist height, chest height growth, green and abundant with many species thriving under their trees: a wood pasture.

These pigs are our ecosystem engineers. Initially it looks a complete mess, but come the spring, after the pigs have been in here for a short, sharp period, and then the ground allowed a long rest, all the different species in the seed bank will come back to life.

I've taken loads of photos to document the progress. We've left a control piece as well – a piece that we're not going to graze with the pigs – which we can compare to the area the pigs have worked over. We're hoping to see a flush of new life in there, loads of diversity, loads of different species of grasses, herbs, legumes. The pigs may bring so much life back into it that we can graze it with the cattle later on.

The rest period is a key part of regenerative agriculture. Whether it be the cows and the mob grazing, pigs in woodland or chickens moving around the farm, rest is key for every single farm. All the nature, all the plants – they all need rest.

I just love the pigs in the woods. Why would we get pigs and put them in a shed? Pigs are meant to be out in woods. These pigs, when you look back at the breed history of these guys, were bred to be in woodland. They are doing regenerative restoration work, and that's what they've been bred for over hundreds of years.

WHY WOULD WE GET PIGS AND PUT THEM IN A SHED?

Lots of people shove them into sheds, feed them ad-lib cereals and concentrates. And, that's why you get a really dry, tough pork chop. I'm not interested in that. I want really nutrient-dense, healthy produce and for them to live their best lives.

It's really important to choose the right breeds for the system that you want to operate. We looked into what breeds work well in woodland and outside, such as Tamworth and Oxford Sandy and Black, which are two breeds that historically haven't been crossbred with Asian pigs, so they still have their hardy traits. Farmers have this misconception that their animals are needy, that they need pampering, so they bring them indoors, which then leads to the massively intensive systems that we have now. The reality is these guys are tougher than what we give them credit for.

We'll probably be sticking at two or three, but we could maybe have five to ten pigs in here. Perhaps a sow that has a couple of litters a year, which we could fatten on the acorns. The only thing I don't like is having to feed them the cereal concentrates. I'd like to find our way into a system where we don't, but at the moment we don't grow any of our own cereals here. If I had arable land with a wheat crop followed by a cover crop, we could maybe graze the cover crop off with pigs. But we've got precious permanent pasture here, and I think it'd be quite damaging if the pigs went through that.

Cattle, gases and carbon

One of the major criticisms of livestock farming is that it contributes to climate change because of livestock emissions: carbon dioxide, methane and nitrous gases. With methane being the major contributor. Through our work with Hannah Jones of the Farm Carbon Toolkit (see Chapter 3) we can show that with all our landscape features (the pasture, hedgerows and woodland) we are actually capturing and storing more greenhouse gases than we emit.

HANNAH: Your 9 km of hedgerows alone are capturing more greenhouse gases (72 t CO_2e) than your livestock are emitting (57 t CO_2e). Your hedgerows are big both in height and depth with mature trees within them. How often are you cutting them?

BEN: Every 3 years. Even if that at times, we're really leaving them as long as we can. We want to maintain that structure to keep that

6.1 Carbon balance, Treveddoe Farm, 2022.

	tco$_2$e	%
EMISSIONS		
Fuels	1.80	2.90
Materials	0.42	0.68
Buildings & infrastructure	1.80	2.91
Livestock	57.68	93.51
Total emissions	61.91	100.00
OFFSET		
Hedgerows	-72.55	68.27
Perennial crops	-2.64	2.48
Permanent wetland	-0.08	0.07
Woodland	-31.01	29.18
Total offset	-106.27	100.00
BALANCE	-44.36	

Table 6.1 Emissions and sequestration, Treveddoe Farm, 2022.

IS ACTUALLY MIND BLOWING HOW POWERFUL SOIL CAN BE.

thick, bushy lower area so we have a habitat for small birds and shelter for the cattle.

HANNAH: Looking at the full set of results, Figure 6.1 shows your 2022 carbon balance. You emitted 61.91 t CO_2e and the farm took in 106.27 t CO_2e. Taking one from the other, the figure is a negative 44.36 t CO_2e, , which means you are removing more greenhouse gases in the atmosphere than you're actually emitting.

BEN: That's brilliant! Such a great marketing tool for us to be able to say we're managing our land such a way that it's sequestering more carbon than the cows are emitting.

HANNAH: And we've not talked about your soils yet.

BEN: We've not done the third year of soil sampling yet, so we haven't yet got data on how much our soils are sequestering.

HANNAH: We can model a projection. Over your 30 ha, sampling down to 0.5 m and assuming a bulk density of 1 g/cm^3 (which is average). Working from a soil organic matter (SOM) content of 8% in 2021 rising to 8.5% for 2023 (which is very achievable with good paddock grazing) the SOM could accumulate just under 800 t CO_2e.

BEN: Is actually mind blowing how powerful soil can be.

Regenerating the next generation

We see the impact the animals can have on the land, so many species depend on livestock in the farmed environment. There's a lot of negativity about livestock. Farming can be shown as cows in sheds, pigs in sheds, chickens in sheds, all of the industrial high-intensity, high-input types of farming, but people don't see what we are doing. I don't think it's communicated as much as it should be, probably because the majority of the food comes from the industrial systems.

If people can see that we can produce food in a different way, with our cows being raised in a way that's enhancing the environment, then they can take an active decision to buy that meat or that veg or those cereals. And then the increased buying power with more and more people choosing to spend their money with farms such as ours is just going to drive regenerative farming so it can have a wider impact.

Education is so important. We'd love to connect with local schools, encouraging the next generation to be regenerative – getting children out on the pastures and in the amazing woodland, having lessons in there, seeing the animals, tasting the produce, cooking the produce, and understanding what it is all about.

I think it's a really exciting position to be in, to have this really good-quality product that's produced in a really regenerative way, which is exactly what the farming industry needs and what the public needs as well. We need to be eating more produce – not just beef but vegetables, cereals, other livestock products – farmed in a way that addresses the nature crisis, the biodiversity crisis and the climate crisis. And I think we have the answer here.

Getting more farmers onboard

I think the way that the world is going, a lot of farmers are considering, or will have to consider, low-input systems like ours to remain profitable and in business: the costs of fossil-fuel-derived inputs are going sky high, and the subsidy system is changing too.

I think the biggest way that I could help farmers to change would be by being vocal about it. Testing and data are important, so when I'm in a conversation with somebody, I can back my system up. For many of the conventional guys, regen ag is still looked at as a bit of a trend, a fad. I need to stand up and say that this can work and here's the proof.

There is no rulebook, no textbook to say do it like this. So farmer-to-farmer learning and farmer-led cluster groups are so important for us to work together to share best practices and to pool the learning from our mistakes. Getting farmers on site to see it happening for themselves has a big part to play. I go to some of the big conferences and events, and they're amazing and give you some ideas, but the motivation is actually

WE NEED TO BE EATING MORE PRODUCE FARMED IN A WAY THAT ADDRESSES THE NATURE CRISIS, THE BIODIVERSITY CRISIS AND THE CLIMATE CRISIS.

going to come from visiting other farmers – like my visit to Cumbria, to the Renisons – to see what they're doing and how they're making it work.

Low points

The most challenging part of our first year was definitely the first full winter: the short days are an absolute killer. It's really hard to get up in the morning, out of bed in the dark, and to go out and around all the animals before I go to my day job at Goss Moor. I've really struggled to do that. I've been late to work quite often. I'm in a very fortunate position that I don't get in trouble for that. I've got the flexibility. I just have to work later in the evenings to make it up. It's really hard to balance it out at the moment. Pushing to improve and expand the farm on the side and having a full-time job, always rushing around and having no time to take a step back and just have a little chat, it all puts a strain on the relationship.

One of the other low points arose from my lack of experience. We were having conversations about supplying an online butchery, and we needed the cattle to reach certain weights at specific times. During January and February, we weighed them, and the weight gains had dropped off a cliff, and when we next weighed them, a lot of them had lost weight. And I was kind of like, 'I'm not a very good farmer. What's happening in my cattle?' I started questioning everything and thinking maybe they should be in a shed. Luckily, I spoke to the guy that owns this online butchery, and he told me that everyone operating a 100% pasture system with the cattle out all year was experiencing the same, and that if overall I was maintaining their weight in the winter that was brilliant.

The film has been a really good journey

The experiences I've gained over this first year have reinforced the fact that we need to do this for nature. I've learned so much and brought so much back here to Treveddoe, like introducing pigs and putting the cows into the woods. Being in the *Six Inches of Soil* film has opened up my mind to what's actually possible on the farm. We're on this planet for a pretty short amount of time and, at the end of the day, if you can leave a legacy behind, if you can have a positive impact, that's huge, hugely important. There's nothing bigger or more important than that. If I can improve the small area of the countryside that I manage, that's a really noble thing to

be able to say. To see what happens in the future, when the impact from the changes we've made shows itself, will be really exciting.

Our goals for the next few years here at Treveddoe are to take nature recovery up to the next level. We've introduced the cows and the pigs, but I really want to start seeing more habitats, like the understory in the woodland, develop. We'd like to introduce new enterprises, such as chickens, and broaden the offering from the farm to include eggs and pasture-raised chicken meat. It is a slow build.

When we look back in a few years' time it should be really cool.

Interlude V

Applying agroforestry in mixed regenerative farming

PRIYA KALIA

The agroforestry practice of silvopasture involves trees being integrated into the same land as livestock (most often cows, sheep, pigs and poultry), delivering direct economic and/or ecological gains. The trees themselves, if managed well, can be used as a secondary crop (for example, for timber, firewood or biomass), which can be marketed or used on farm to reduce costs. Trees also produce a broad range of agroecological benefits. Trees along a field boundary provide a suitable microclimate in their shade. This can decrease heat and cold stress in ruminants while increasing productivity. They can decrease soil erosion and help land become more resilient to climate change. Trees can also enable access to vehicles by drying out boggy land and improving biodiversity.[1]

The Agroforestry Show, the UK's first dedicated event co-organised by the Soil Association and the Woodland Trust took place in September 2023. Farmers, foresters, researchers, environmentalists and policy makers came together to share insights and advice on how to help farm businesses

SILVOPASTURE SYSTEM		EXAMPLES
Trees within livestock pastures	Boundary tree systems	Shelterbelts, riverside planting and hedgerows
	Regularly spaced tree systems	Grazed orchards, row systems, clump systems
Livestock within woodland systems	Woodland grazing	Pannage systems, silvopoultry, parklands

Table V.1 The three main types of silvopasture systems.[3]

Soil Association

benefit from trees and established a consensus that trees are key to ensuring food production while tackling climate change and biodiversity loss.[2]

Nikki Yoxall, Pasture for Life

Nikki Yoxall *is a new entrant farmer. She runs Grampian Graziers in Aberdeenshire with her husband James, across multiple holdings, where they run native breed cattle. They are farming regeneratively, using a combination of mob grazing and silvopasture.*

'The biomimicry offered by silvopasture, creating an ecological foundation that mirrors more diverse communities as found beyond the boundaries of many conventional farms, provides enhanced services such as soil health and conservation, carbon storage and increased biodiversity', says Nikki.

'Agroforestry and its subcomponents of silvoarable and silvopasture shift the focus away from food being the primary service of agroecosystems and instead promote a wider range of services that provide more stability and resilience for the system. In addition, by integrating trees into an agroecosystem, there are direct ecosystem services to animal health and welfare, leading to increased farm productivity and the potential for reduced input costs', Nikki adds.

In a future where extreme weather events will impact livestock management decisions alongside a need for more environmentally sustainable farming practices to maintain food production without further degrading soils or plant and animal communities, Nikki thinks that agroforestry will likely become increasingly important as a solution. She concludes: 'The design of complex planting schemes, or even adding single trees scattered throughout farmland can bring multiple benefits to the resilience and persistence of an agroecosystem and the services it provides.'

Matt Chatfield

Matt, *a Cornish sheep farmer and 2022 runner-up,* Farmer's Guardian *'Sheep Farmer of the Year',* talks about his pioneering transition to silvopasture.

'I thought to myself, if I can do it, anybody can. Before that, I'd never even used a chainsaw. I had no idea about woods. It was basically in December, so I didn't even know what the trees were. So, my whole philosophy is you need light in the canopy above all, and in terms of all farming, you're essentially feeding worms in the soil.'

Matt's affinity to earthworms has only grown over the years: 'I now consider myself now like a worm farmer than like a sheep farmer. If I get the worms right, dung beetles are incredibly important, but the whole system is designed. I need to feed the worms. They can bring organic matter into the soil.'

The other pleasant and important surprise Matt witnessed as he practised silvopasture was the absolute 'explosion' in biodiversity. 'Previously, we had about six plants coming from the seed bank. Three years later, we had about 100 plants springing up. I haven't put anything in there – the explosion in biodiversity is just absolutely ridiculous.'

Matt fattens sheep no longer fit for breeding and moderates the time the sheep spend in a section of wood depending on the size of his flock. So, if he's got 80 sheep, he puts them in for 2 days. With a flock of 160 sheep, he'd only put them in the same area for just 1 day. 'Just keep them moving around, basically,' he advises.

Coppicing can be very beneficial in a silvopasture system, but at first, it might seem a bit drastic to cut down so much wood. Matt describes his experience: 'When you coppice, when you first coppice, you think you're killing it because I've never done it. And then when you see it grow, that's quite amazing.'

With all these changes, one might question how silvopasture might have

influenced the quality of the mutton produced. Matt has had positive responses to his silvopasture-raised sheep. 'There's a chef called Jeremy Chan, whose London restaurant has two Michelin stars. He said, "I think this already compares to an Iberico pig." The flavour is in there. All I'm doing is adding this interesting fat flavour which I think comes from the biodiversity of what the sheep are eating.'

Q&A Jake Locke, Founder & CEO Silvasheep

The Silvasheep project is developing a regenerative agroforestry system that unifies food security, farming and the environment. Jake Locke, CEO, talks about his business model and how it incorporates silvopasture and silvoarable.

Q: Can you tell me a bit about your background and about Silvasheep? How does it employ silvopasture and silvoarable methods in your work?
A: I have a background in branding, an interest in active outdoor activities and healthy food. I had read a book some years ago by Olivia Mills called *Practical Sheep Dairying*. These elements came together during a walk in fields and woods during the COVID lockdown in 2020 and the idea for agroforestry sheep's milk yoghurt

was born. I met regenerative farmer Tim Coates, after Groundswell in 2021, and land asset manager and farming consultant, Andrew Long, at a Defra agroforestry workshop shortly after. Sharing a vision, we became Silvasheep co-founders and started a pilot agroforestry operation at Tim's Cotswolds farm in 2022.

Agroforestry can offer benefits to the soil, biodiversity and the animal in terms of shade and shelter and a better nutritional diet. On the side of product quality and health, agroforestry allows us the opportunity to explore beneficial properties contained in certain tree species' bark and leaves and whether these phytonutrients can be transferred to the milk to enhance nutritional benefits for the consumer.

Silvasheep's main product is a natural strained yoghurt made with milk from sheep grazed outdoors in a combined silvopasture and silvoarable environment. Secondary agroforestry products from the Silvasheep system include wool (for environmental use and clothing), meat, fruit, nuts, botanicals and timber.

We are focused on preserving provenance and rural communities to help ensure UK food security. Routes to market will include local food supply chains, for example, farm shops and/or direct-to-consumer (DTC) platforms, as well as retail outlets including supermarkets.

Q: Can you tell us more about your approach, and how it ties in with regenerative farming?
A: Our approach is flexible. For example, in the case of a mutually beneficial arrangement with a farmer, for granting grazing access to Silvasheep's agroforestry system a landowner receives an improvement of the land and the environment. This natural capital can now be realised to provide ecosystem services in the form of carbon offset, clean air, improved soil health, biodiversity gain, and reduced drought and flooding. Although, currently in flux and often confusing, the demand for ecosystem services is expected to grow considerably in coming years, therefore, the Silvasheep model offers an attractive commercially viable alternative to traditional grazing rent arrangements. We offer joint ventures with Silvasheep in which product costs and income streams are shared according to agreed splits.

In terms of applying the system in the field it is important to look at things holistically – incorporating natural capital as well as animal welfare considerations. Alongside the improved nutrition an agroforestry system can provide for grazing animals, the shade, shelter and perceived protection trees and shrubs provide is likely to cause an animal to be calmer and, in turn, have more efficient rumination than a more stressed intensively farmed animal

might demonstrate. A healthy animal producing better quality food for consumption is the main aim, but also a system that produces less greenhouse gas (GHG) emissions will be valuable in achieving better natural capital accounts that are now encouraged in farming systems.

Q: How do you work with farmers to develop silvopasture and silvoarable methods that work for a particular piece of land?
A: We don't pretend to be the experts in a particular farmer's land – the farmer will be better placed to understand the combination of local land topography, land culture and environment. Of course, certain Silvasheep elements are transcribable, but they often need to be adapted to each individual situation. We're keen to avoid a one size fits all approach as there are too many complex nuances in local ecosystems. We would take this further and say we would promote regional and local differences perhaps in subtle variations in taste and texture, as long as the Silvasheep approach, quality and nutritional density remains consistent.

Q: How are you building trust with farmers, to get them on board with this relatively new concept in farming?
A: Well, having a third-generation farmer as a co-founder helps, as does putting in the work ourselves in the

field in the Silvasheep pilot in the Cotswolds. We're also putting our money where our mouth is and giving research the respect in needs. Our vision is steered by an authentic voice. Our belief is that it's all about the health of the land, the animal and the people involved – get that right then it will have real value, people will come, and it will be commercially viable. What we're not about is additives, hype, or squeezing the farmer's purse.

Ultimately, we aim to build a strong brand that's powerful enough to be commercially independent of subsidies, whether or not current and future payments are actually taken. We think this will be attractive to farmers. We don't doubt that there is work to be done to gain trust within a sector in transition, but we are encouraged by the positive discussions and relationships we have had and continue to have every week.

As well as being flexible, our business model aims to be collaborative and inclusive by nature and aware of the local context in which it operates. For example, as the brand develops then so can investment in local communities in terms of job creation both on farm and in the local supply chain, which might include local shops, pubs and restaurants. With the development of DTC platforms (like Ooooby, for example) people might eat out, love what they've eaten, see it is available online produced at their local farm and order it.

We see ourselves as proactive and highly supportive and adaptive agricultural entrepreneurs – Draconian rules just won't work. We like the word 'regenerative' because it means the metamorphosis of something into something better than it was before – what's not to like? Whether 'regenerative farming' needs to be stamped and watermarked and burdened with rules and regulations, I'm not so sure. At Silvasheep, we try to incorporate certain aspects of silvopasture and silvoarable principles into our work but also maintain an open mind and a spirit of careful experimentation, observation and measurement. This is also why we often call the Silvasheep project an 'expedition'.

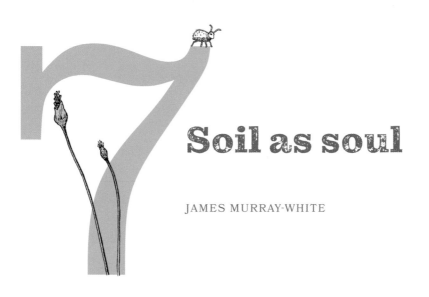

Soil as soul

JAMES MURRAY-WHITE

Soil is a source of life and our bodies are soil transformed.

SATISH KUMAR

Without exception, all of the new entrants into farming, established farmers, experts and ecological philosophers we've talked to during the course of making this film, talk about their connection to soil. For some, it's a place of work, for others, it's a place of belonging and home. For most, it represents everything that they do and are. And this, in essence, is what I mean by soil as soul.

> Both philosophically as well as ecologically, putting your hands in
> the soil, means you know who you are.
>
> VANDANA SHIVA[1]

Vandana Shiva is one of the great heroes of our age; an activist, scholar, seed saver and much more, she is hugely inspirational in seeking to show that we can have autonomy in our well-being, and that we need to stand up against the industrial mindset that seeks to destroy the earth's resources – big pharma, big food, big logging companies, big genetically modified and chemical agriculture. Considered together with activist-philosopher Satish Kumar, founder of Schumacher College, I feel blessed that we have these two inspiring figures who both straddle different continents and societies, and who are unafraid to challenge systems and hierarchies as well as provide hope.

Soil communities

Six Inches of Soil is trying to give a flavour of the human community that cares passionately about the health of food, and the soil that nurtures it. Our early research phase (of finding the what and who) and then the filming phase (with the principal characters and the experts that we wanted to go deeper with) took us around the country meeting amazing people.

As a growing team we have explored angles, issues and side debates in early conversations with farmer George Young at Fobbings Farm in Essex, on a private tour of the Holkham Estate in Norfolk with conservation expert Jake Fiennes, and with the many incredible experts like Dee Woods, building community care and resilience in London, and permaculture farmer, Hannah Thoroughgood in Lincolnshire (now faced with compulsory purchase to make way for a reservoir). We've met seemingly unreconcilable opposites – journalist and climate activist George Monbiot

WITHOUT EXCEPTION, EVERYONE WE'VE TALKED TO
DURING THE COURSE OF MAKING THIS FILM, TALK
ABOUT THEIR CONNECTION TO SOIL.

and his call (in *Regenesis*²) to move to feeding populations protein created from artificial lab-grown bacteria and Frome-based, ex-academic and small-scale farmer Chris Smaje and his resilient retort *Saying No to a Farm-Free Future*³ – within which there is still common ground to be found: celebration and recognition of the importance of soil.

Some of these experts didn't make it into the final film, purely for the reasons of time and having far too much material, so I wanted to mention them here, right at the top, to thank them for their work and efforts to promote land care in the ways they do.

Jake Fiennes, in his influential book, *Land Healer*,⁴ powerfully argues that traditions of care within multiple scales of farming need to be reclaimed (as he is doing so profoundly on the Holkham Estate in Norfolk) and the wildlife and multitudes of flora will return, and the biodiverse ecosystems will flourish. Asked if he'd like to provide a quote for this chapter, he shot back with:

> Our food is provided for us by one of our most important natural capital assets, SOIL. For without it there is no food on our tables, it sustains all of us alongside the wonders that are nature. It is our responsibility to ensure that we get it in a condition that is fit for future generations!

Anne Biklé (who gave a powerful keynote lecture at Groundswell in 2023) and her husband and writing partner, David R. Montgomery, whose book, *What Your Food Ate*⁵ truly turns our understanding upon its head and simply makes us reconsider our relationship with food and soil. She talks directly of 'spiritual renewal or revival, that is hopeful and makes me feel connected to land and people again', through regenerative farming practices, that create a 'biological bazaar', balancing micronutrients, phytochemical, a fat balance and microbial metabolites, within the circular system of soil and food.

Of course, we don't have to start with this top-heavy human layer of the peoples of the soil, we could start the discussion from the soil up. Or even downwards, or even scanning horizontally, through layers of matter and mycorrhizal material. We need to get immersive sometimes simply to know place and purpose, or as Vandana Shiva says, to know ourselves at all! We hope we've managed to put soil front and centre of the film, and have only just started to scratch the surface of the footage that is emerging through new developments in camera and lens technology – microscopy reaching new gradations of ability to see the tiniest forms of life at work within cultures and depths. It is crucial to keep in (humankind) mind that a practice of de-colonising our control of soil is to always attempt to give voice to the more than human. Here, now, let's realise this effort to give voice to soil – all the life within and the form of earthy soil itself.

Recent popularisations of mushroom culture and how the mycelial networks abound within soil, using it as their substrate to thrive by biologist Merlin Sheldrake, anthropologist Monica Gagliano and 'mother tree' expert Suzanne Simard, amongst others, show that these knowledges exist and need disseminating, across all platforms, to enrich public understanding of soil. Anne Biklé writes of the rhizosphere (the region of soil in the vicinity of plant roots in which the chemistry and microbiology is influenced by their growth, respiration, and nutrient exchange) as 'the truly wild and alive place'.

Reading the obituary this week of renowned transplant surgeon, Sir Roy Calne, I was delighted to read that his work was only possible through the research led by the Swiss chemical company Sandoz, which collected soil samples that contained the fungus that was developed into the immunosuppressant drug cyclosporine. All life – and mycorrhizal immunosuppressants – are in soil.

Maybe such deeper knowledges will enhance and influence this growing movement of care for soil? Ultimately, many of us at death return to soil. For those that choose burial, it is our final resting place. Given our dependence upon soil – our standing upon it, reliance upon the food grown within it or nurtured from it, and the likelihood that we will be lowered into it at our end or scattered to the winds above it, soil is ultimately our human origin story.

When our connection and understanding becomes as stark and as simple as this, our effort to care, and wonder, and maybe study of it, all fall into place.

Social regeneration

Food is naturally interwoven through the stories we have brought to screen, from the animals that Ben Thomas cares for at his farm in Bodmin and takes through the circle of birth to death for meat production, through to the grains on Anna and Andrew's Lincolnshire land, to the range of vegetables bursting with life and flavour at Sweetpea Market Garden, Adrienne's beautiful land at Caxton in Cambridgeshire. I've volunteered at Sweetpea, sown seeds there by hand, manured the squash beds and helped harvest chard, spinach, many varieties of tomatoes and more, so can absolutely testify to the life-giving properties and textures of vegetables from that soil. The land is heavy clay, so as well as lightly tilling it, following min-till principles, Adrienne has brought in well-rotted compost from a nearby chicken farm. That farm specialises in re-purposing chicken manure in a range of composts and liquid plant feeds, most of which I've used on the allotment as well.

Adrienne says of her reasons to be a grower:

> I have a love of growing food and the resolve that we need to be producing food locally and in an environmentally sensitive way for our future resilience. My hope is that it's possible to create a thriving local business that will weather the storms through community support and in turn support the more vulnerable members of our community.

Ben writes of his farming passion:

> I really like the idea that everything we produce on our farm that leaves the farm gate, has eaten only food that has come from the

223

farm, making it truly resilient and not reliant on external or global markets. I love not worrying about the price of fertiliser, the price of red diesel or feed prices for cattle. What is giving me encouragement to keep going with this farming method, is that you become more in touch, or in tune with the seasons and therefore nature. I take happiness in hearing the dawn chorus all spring after a bleak winter, in seeing all the invertebrates absolutely thriving in the long grass, the cuckoos, the abundance of earthworms and the flocks of small birds following the cattle during bale grazing in the winter. It gives me pride seeing and hearing all these things, and it is what humans have become disconnected from in the last two generations. I do not want to be disconnected from our natural world.

How food is processed, and how it comes to market, and indeed what those markets are, is part of the wider story of the value of real food, grown and nurtured within 'good' soil, that we are exploring within the film, with the hope of really being part of the growing movement for change.

There is the possibility of social regeneration if we really take care to re-examine as a functioning society how we eat, and how we come to food. Thinking on a personal individual level, and then on a family or household scale, about our food preferences – what do we want it to look like, to smell like, how we receive it and most of all how we choose it – it's a crucial tool in the steps to creating change. Once we can operate and examine at this micro level, then we can scale up to our communities and the wider sense of society, on national and international levels. Start with ourselves, then our village.

One of our key inspirational characters, Dee Woods, award-winning cook, community food educator, urban agriculturalist, co-editor of 'A People's Food Policy',[6] and co-founder of Granville Community Kitchen (London), has been doing this for many years, and movingly describes herself:

> The Earth is my mother, I am of the soil and I will return to the soil. I describe myself as a land elder, weaver and connector and knowledge broker, building equity into systems means I do many different things and wear many different hats but all are connected to cooking and community: teaching children and young people how to cook, to grow food, to build and localise food systems, to conduct research and political advocacy.

It's our choice then, as readers, as writers, filmmakers, activists, farmers, people of place, community, opinion and habit, and definitely as eaters, to engage with the soil below our feet. Whether that is following the links to the various campaigns we as a film team promote, or to do what the farmers and philosophers advise, and to buy a regular vegetable/meat box from your local community supported agriculture project, we can and must be a deeper more connected part of this system.

Saving these 'Six Inches of Soil'

A key centre for holistic and regenerative education and practice in the UK is Schumacher College, on the 1,200 acre (486 ha) Dartington Estate outside Totnes – once the pioneering home of organic farming techniques put in place by the Elmhirsts, the college was founded by Satish Kumar in 1991, as part of the Resurgence Trust. I have been fortunate to attend several short courses there, heard Satish speak many times and given that one of his books is titled *Soil Soul Society*[7] it felt inevitable that Satish should be part of this project. One of the film's impact producers, Elsa Kent, a farmer's daughter from south west Devon, a filmmaker and ecological education activist, has recently graduated from Schumacher:

> Schumacher College facilitated an exploration not only into ecology, regenerative agriculture and deeply transformative practice, but into myself, and my spiritual connection to Gaia.

> This is something I always knew was there, but I didn't really have the tools to express it, or allow it to guide me through life. At Schumacher I was surrounded by a thriving community of beautiful souls who love the soil as themselves, who see the woods, birds and

earthworms as sacred kin, and who know how to love the world so profoundly that it guides each day here on earth.

This may sound quite 'woo woo' but my experience of Schumacher College and these people was that they are the most proactive, engaged and thoughtful citizens I have met so far. They have committed their lives to taking care of people and planet, and every day they throw themselves into the movement with zest, vibrancy and hope. I wonder if we cultivate this highly instinctive spiritual connection to the land and Gaia, whether this enables us to have more resilience in our activism? When my actions are fuelled by love and connection, not anger or frustration, I feel stronger, and better enabled to keep up the mission into the future.

As dear Satish explains, this is how he has been able to live such a long and vibrant life of purpose: 'Do not worry,' he says – 'just act!' Active hope now forms the foundation of my approach to this incredible movement, and I could not be more grateful to Schumacher for helping me to find this path.

Satish Kumar

In the interview for the film, Satish told us:

I come from India and Indian culture believes that the soil is sacred. Soil is a source of life and our bodies are soil transformed. All our food is soil transformed. All the buildings we build with wood and bricks. They are soil transformed – soil is the source of life. Soil in the modern world, in the industrialised way of thinking is seen as a resource for the economy.

My message to farmers is, please don't see food as a commodity. Please don't see land as a commodity. Food is sacred. Land is sacred. Please produce food to feed people. And please produce food with love and compassion and not as a mechanised process to make money or make profit. In our modern mechanistic world view we only recognise things which can be measured. What cannot be measured does not exist. I think this worldview needs to change.

We are human body as well as we are human spirit. You can measure the body, but you cannot measure spirit. The spirit is where our compassion, our generosity, our kindness, our relationship, our friendship, our love, lives. We have to create beauty in our lives. We should not have anything which is not beautiful because without beauty, soul, human spirit will die.

At the moment humanity is facing many big challenges: climate change, diminishing biodiversity, pollution and waste. It is our responsibility to stand up and speak truth to power. It is our responsibility to do our best in the service of humanity and in the service of our precious planet Earth. Because we are microcosm of macrocosm. We are made of earth, fire, water, we are made of consciousness, intelligence and imagination. And therefore we have to take responsibility to look after the earth, air, fire and water. Take responsibility to look after our fellow human beings. We have to take responsibility to look after our climate, take responsibility to look after our animals, and all the living beings, and especially the soil!

227

After the film première at the Oxford Real Farming Conference this January, Satish responded to watching it by saying: 'The word for soil in Latin is *humus*, and the etymological root for humans is also *humus*. It is the same, so human beings are literally soil beings. We are all made of soil! This six inches of soil has literally made us! Our bodies are soil transformed.'

Climate and food academic, Dr Lucy Michaels, part of the Six Inches research team, is an anthropologist interested in food and soil and their social meanings. Her passion is 'soil voices', and linking human connection to land and place with the voices of those that grow food locally, on any scale, and how getting our hands in soil boosts immunity and well-being. She expands on both Satish and Anne Biklé's words:

> Soils are not simply 'territory', nor a bio-physico-chemical compound. They are not just a 'resource'. Soils are a multispecies living world made of relationships that include the humans who care for them. Healthy soils rely on thousands of living organisms interacting with each other and with abiotic matter: feeding each other, being fed, living and dying. Something that we tried to convey in the film, was the way in which healthy soil relationships beneath the ground, reflect healthy agroecological systems above ground too. How we eat and feed each other as humans should primarily be based on our relationships with each other and with the land, rather than simply a disconnected transaction. If we stop and listen, the soils have so much to teach us.

Halfway through this project, a book landed on my desk, sent to review. I loved the title, although busyness meant it got put aside, and yet I knew that it would resonate with me, and what I hope we are trying to underpin throughout the film, and indeed this accompanying book.

Scott Chaskey's *Soil and Soul*[8] is a vibrant and bold statement of commitment to soil. Chaskey is a unique combination of farmer and poet, a man of the soil who has reaffirmed his belief by pioneering organic certification (in the US), and pushing for it in China too. I resonate with his writing on how it is crucial to recognise ancestry and indigenous belonging of and to a place. This reversal of colonial practices echoes particularly the power of Vandana Shiva's words on how oppression of peoples, as well as ultimately soil, always backfires. And as Chaskey so humorously observes, 'Often we travel to return to then truly know our place and our rooted living soil.'

A final shovelful

We then started, 3 years ago, brainstorming ideas and storylines for a bigger piece on this shift towards agroecology, it felt like a vast pool of people, ideas, theory, as well as good and bad experiences were out there, awaiting us rocking up to ask them questions and point cameras. This is a beautiful, almost sacred, meeting of peoples, stories, ideas and potential – truly crossing cultural divides that separate humans through class, money, land access and all the other terrible fault lines that haunt us.

In a similar way to how all those we've met on this regenerative journey tend the soil with love and care, we as storytellers use their stories, and the technologies we have at our disposal, from cameras to editing software, with equal tender care in creating a visual journey to craft a deeper narrative.

In her recent book that breathes new life into old mythological stories, *The Flowering Wand*,[9] American writer, educator and poet-activist Sophie Strand exhorts us towards the vegetal world, as a way back to our senses again: 'We don't need a grand spiritual awakening in order to come back into this Edenic consciousness. The garden is just beneath our feet.'

Her work builds to a resonant conclusion, as if reaching down through composting layers, feeling decomposing leaves, peelings, the slippery touch of earthworms and red brandlings, ants, gnats, protozoa, springtails until our hands find in the lower levels words pulsing and coalescing into stories that emerge through the rich peaty matter:

> These days I'm much more interested in the dirt, so I'm offering a new gnostic maxim: ensoulment is enrolment. Soul is soil.

Interlude VI
Enterprise stacking

PRIYA KALIA

Stacking aims to maximise output by running several enterprises alongside each other. It's a form of holistic management that puts regenerative farming at the heart of a commercially viable farming business, 'layering' businesses on top of each other that connect and benefit each other. The community it builds can reduce the loneliness farmers can feel by providing a closer, local peer-support system in which entrepreneurial successes can be shared and celebrated together. Challenges can be overcome as a group.

There are some good examples of stacking in the main chapters. Adrienne and Tom bringing in a local shepherd and his flying flock (Chapter 5) and Ben's introduction of pigs and his plans for a mobile hen coop (Chapter 6). Of course, Adrienne herself is a stacked enterprise hosted by Tom, her landlord farmer, to the benefit of both of them.

The Pink Pig

Anna and her parents (Chapter 4) have introduced several enterprises over the years ranging in scale from their recent partnership with a local beekeeper to the much larger diversification centred on the Pink Pig Farm attraction.

The family's diversification into other businesses started when Sally Jackson (Anna's mother) started selling eggs and cakes from a trestle table at the end of the farm drive over 20 years ago. From this, she extended her business with a chicken coop, restaurant, and farm shop that, over time, continued to expand.

Today, the chickens and farm shop are no longer running, replaced by a successful restaurant, celebration rooms and an indoor and outdoor farm park that brings in and engages different members of the local community. A ground-source heat pump and solar energy heat the indoor spaces. School groups also come to the farm to learn about where their food comes from. The farm also boasts an aerial trail, with a high ropes course for kids and adults, in collaboration with an external partner.

Anna and a business partner await permission from the planning inspectorate to run a doggy day care, providing her with a second income to supplement her part-time farming income. Anna is working with an entrepreneur with a background in fashion to use her sheep's wool to make advanced, breathable and warm sportswear material interwoven with nettle and hemp. Wovenbeyond's natural and fully traceable yarns are made with fibres grown through farming methods that operate in synergy with their surrounding ecosystems and actively engage in restoring nature – traceable, agroecological wool.

The farm also hosts an annual food and music festival in May. Anna is also looking for other ways in which to diversify on the farm, including potential collaborations with a local abattoir to process her lamb and possibly set up an online meat business.

'People think running a business is profitable; however, in farming, although you won't earn millions, you'll make money and have the best time', concludes Anna.

Diversification through enterprise stacking is an essential way to make it all work, including lifestyle, livelihood, environmental and, importantly, financial considerations. There follow three case studies of successful stacking enterprises in what have now grown to quite large concerns.

Andy Gray, Elston Farm, Devon

Andy was born into a farming family in Devon with hundreds of years of farming and cider-making heritage. Soon after agricultural college he had the opportunity to take the helm of a failing chicken farm and slaughterhouse. Andy decided to restructure it to create a multispecies meat business – the start of his entrepreneurial career. He eventually acquired the business, then later the farm surrounding it – Elston Farm – running it as a mixed arable and livestock farm.

In terms of stacking enterprises, Andy's meat business, which has been running since 1955, supplies hotels, restaurants and butchers across Devon and Cornwall. He also processes wild shot deer and other types of game for the same customers. Disliking waste, Andy and his colleagues set up a dog food business utilising trimmings and offal that had no other outlet.

Andy also got into selling meat boxes as an inevitable result of the COVID-19 pandemic, when his hotel, restaurant and high street sales plummeted. His team set up a website in a week and started selling UK-wide.

After contacting the Sustainable Food Trust (SFT) for advice, Andy was asked to licence the UK's only mobile slaughterhouse, which he has trialled on behalf of the SFT's Chair, Lady

Freshly planted trees, pollen, wildflower mix and wind turbines at Elston Farm.

Jane Parker. 'Small slaughterhouses are critical for regen farming', he says. The Isles of Scilly, and other islands are interested in the concept to service livestock farms on their islands, where otherwise livestock would have to be transported to the mainland at a high cost. Livestock are critical for maintaining habitats and nature recovery on many islands.

Aside from farming, Andy's farm generates a lot of its own energy from wind turbines and solar panels, which he feels a duty to do as a meat producer to offset the carbon footprint of refrigeration. It helps with their financial sustainability. The energy also completely meets the needs of the tenanted properties on the farm, which are also a source of income. Elston is also rich in a rare stone that can be used to mend heritage walls. Andy was able to acquire a quarrying licence and today operates the site with fairly low overheads. Locally quarried stone is a low-carbon alternative to concrete.

Andy's got a couple of start-ups on the horizon as well. One is a collaboration trying to process bones and skins into fertiliser in a safe and approved process. 'The theory is that grass can be turned into a bullock, so we can likewise turn the by-products from the bullock back into grass – a mini circular economy.' Another involves assisting a University College London researcher who has developed a means of photosynthesising using micro-organisms to produce a material that offers the compressive strength of bricks. Currently in the development phase, the trial and scaled-up manufacturing of the product will start at Elston.

In addition to his entrepreneurial activities, Elston is also host to a series of different scientific studies, in collaboration with Rothamsted Research and others. With various commercial enterprises, combined with support for scientific research, Elston is a great example of enterprise stacking on a farm.

Tim May, Kingsclere Estates, Hampshire

Tim is a fourth-generation regenerative farmer at the family-owned 2,500 acre (1012 ha) Kingsclere Estates in the Hampshire Downs. Tim undertook a Nuffield Farming Scholarship in 2011 during which he travelled to the US, Brazil and Africa.[1] He mixed with farming's future leaders, innovators and disruptors and had the opportunity to step back and reflect on his business. He realised that the monoculture of arable cropping on his farm was not financially or environmentally sustainable.

Tim (right) with Oliver Chedgey (left) of Roaming Dairy Ltd, with the mobile dairy in the background.

Returning to his farm, Tim felt that grass would be important as part of the rotation in the growing system to improve productivity. 'We had some knowledge about grass in the business, when we had a pig farm and a dairy, although these businesses closed down in 2002 due to financial pressures.'

In 2013, Tim invested in fertility farming, seeding herbal leys, having a full year of crop rotation and reintroducing livestock to the farm. He started with sheep, buying a substantial flock of 1800 ewes from a retiring local farmer. Tim quickly moved on to cattle. At that time he started to experience significant mental health issues. 'I thought, what am I missing here? And then I realised, I needed to share the management of the land.' Tim introduced a share-farming model, with people running their own businesses among his. He realised the more diverse business activities one has on the farm the more upcycling and benefits everyone experienced. This is where the idea of his regular 'Pitch Up' sessions came from, which have gained Tim and Kingsclere Estates recognition for their open-minded, proactive approach.

One entrepreneur came in and set up a roaming dairy, working with the grass system and making use of mobile infrastructure. They hosted a 1,200 bird egg business for a time. Kingsclere Estates also hosts a pet food company, which picks leaves from the farm's hedgerows and dries them, providing mainly London-based rabbit owners with regenerative hay and other food. His wife also runs her counselling business, with a team of 15 counsellors, on the farm.

In the main, people coming to start their businesses at Kingsclere Estates often are looking for three main opportunities: **producing**, creating raw materials; **processing**, which adds value to raw materials from the farm or brings raw materials in and convert them on the farm; **by-products**, which can be useful on the farm and make it more circular. One example of this would be to have a cheese maker producing whey as a by-product, a useful fertiliser for the land or as feed for chickens or pigs. We also offer marketing for those individuals or businesses taking sustainable products to market within the local area, either business to business (B2B) or direct to the consumer (business to consumer, B2C).

His next plans include a food enterprise hub that can help businesses with processing. 'We want to create a space where people can explore creating new food, fibre and energy products within the local area, and to support these businesses for the first three years and help them grow.' Tim states. 'At the core, I tend to follow the three pillars of sustainability: people, planet, profit.'

Westmorland Family

The Dunning Family have run their Cumbrian hill farm for generations. For the past 52 years, the family business has included the successful motorway service stations Tebay Services, Gloucester Services and Cairn Lodge Services. The service station enterprise was originally founded by John and Barbara Dunning in 1972 when the couple realised a commercial opportunity that was literally on their front doorstep.

'In the late 1960s, the M6 was built through our family's farm. My dad, John, was a natural entrepreneur, and was keen to expand other interests alongside agriculture', shares Sarah Dunning. 'My parents bid to buy back their land, which the government had acquired on compulsory purchase, 52 years later we have over 4 million travellers a year coming through our services.'

Tebay, and more recently Gloucester Services, offer delicious, homemade food, drinks and desserts and a farm shop. They source much of the food and products from local suppliers and farmers within a 30-mile radius. It's a stark contrast from the ultra-high processed fast food and supermarket chains that dominate many of the UK's service stations.

'To create a sustainable future, we couldn't just rely on farming alone. The fact is that upland farming doesn't provide a livelihood one can easily live on', Sarah explains. 'My parents did what was natural for them – to source locally and make their own food from scratch for travellers. And their model worked.'

Today, daughters Sarah and Jane run the business and the farm and have steadfastly stuck to the roots of their parents' diversified business, focusing on the local economy to make a local impact. This approach has been successfully combined with a shift towards improving soil health and productive land use on the farm. Sarah and Jane have consistently sought to partner with local food producers. The joy of their model, Sarah explains, is to try and find small businesses and ambitious, talented entrepreneurs and to work with them, not as a standard business relationship, but as partners from the start: 'We work to the motto of "You grow, we grow"; we try to help each other all the way along. We prioritise local because we want a model that is going back into the local economy.'

Tebay Services Southbound on the M6.

In the Westmorland Family farm shops, the aim is to work with as many small craft producers as they can find. Sarah explains: 'Our priority is: quality first, followed by local. We would never select a local producer that wasn't high quality to supply our shops.' In the Tebay area, they work with 70 local producers, building their businesses together. 'In Gloucestershire, we have a much more diverse, rich producer community compared to the Cumbrian uplands', Sarah continues. Gloucester Services sources its food, drink and products from 130 local producers within a 30-mile radius. The family also have residential properties on their Cumbrian farm, with holiday cottages and long-term lets.

Considering the future of their business, the biggest transformation that Sarah can see is a switch from petrol sales at motorway services to electric vehicle charging sales (EV): 'The implication of this is that we need a lot more power on site. It is difficult to get that amount of power directly from the grid.' The family hope to install solar panels in the land surrounding Tebay Services, to provide this electricity once suitable planning permission is received. 'Sustainable energy can be a real opportunity for farmers, if the energy generation is close to the required usage.'

In addition to their many businesses and commercial partnerships, Tebay Services Northbound also hosts a horticultural charity called Growing Well, which has their site rent-free. This charity works with local residents with mental illness to grow fruit and vegetables, effectively providing a form of therapy.

How do we build a sustainable future for food and farming?

CLAIRE MACKENZIE, COLIN RAMSAY,
MOLLY FOSTER & LUCY MICHAELS

With contributions from

MIKE BERNERS-LEE, HENRY DIMBLEBY,
VICKI HIRD, HANNAH JONES, SATISH KUMAR,
TIM LANG, NICOLE MASTERS, JOSIAH
MELDRUM, JOHN PAWSEY, IAN WILKINSON
& DEE WOODS

We've posed a series of questions to our experts and collated their answers around topics so that you can explore their opinions on big-picture solutions for food, farming, agroecology and society at large. This is not a shopping list of solutions, and many of our experts are talking about fundamental value shifts for society. The details of how we care for the land and the soil can be found earlier in the book, such as in Chapter 3; here we are taking one step back and surveying the wider landscape. What is crucial is that the great work of the farmers featured in *Six Inches of Soil*, and many others like them, does not and cannot happen in a vacuum … change might be in the soil, but we also need change in the wider system.

CHANGE MIGHT BE IN THE SOIL, BUT WE ALSO NEED
CHANGE IN THE WIDER SYSTEM.

CLAIRE MACKENZIE: We've been on such a long and fantastic journey with this film and we've learned so much from the farming and food community and the organisations we have partnered with. Nature is of course the greatest teacher of all!

COLIN RAMSAY: Physically being on those farms – walking the land, speaking to the farmers, absorbing all the natural energy, seeing how you can take a seed and see what it grows into – all of that opened up something inside me, noticing what's possible in nature. It's a very powerful transformation and now I can't put the genie back in the bottle.

I just know that this works. Maybe it is actually a silver bullet, if it's done right, and if it's done with support and good intentions. Change is possible and we can turn the nature, food, and health crisis around.

CM: I maybe want to call it a gold bullet. In my mind a silver bullet is a quick thing, something people are investing in quickly, but this is long-term and holistic. The endless money that is being invested in high tech solutions is worrying as this is how we have always tried to extract ourselves from problems. We need to slow down, we need to work as communities and learn more from the soil, land and nature. With the ever-changing climate it is hard to predict what is next but we know protecting our soil enables us to farm for the future. You've got to go on that journey, but governments, corporations, us as humans, we've become more and more now, now, now!

CR: We know what the solutions are. We've heard from experts like Vicki Hird and Tim Lang who've been working on this for years. We just need to act on it.

So, what can you do as an individual? You can go on a farm walk. You can buy once a week from your local butcher. You can start growing herbs in a window box. Begin the transformation in your own back garden.

To connect to the land, you have to spend time on farms, you really do.[1] You've got to spend time putting your hands in the soil, growing things, speaking to farmers about the challenges they face. And then it puts a whole new perspective on going into the sterile, corporate environment of the supermarket, looking at everything wrapped in plastic and just buying cheap stuff that's come from halfway round the world.

CM: I think I've always had a vision with this film that people would want to put their hands in the soil having seen it. I was quite influenced by Lucy Michaels,[2] who was studying soil and meaning, and ecological belonging, looking at how people could re-root themselves by growing things. My deepest memories take place on my Grandma and Grand-pa's farm in Wales. I was on their dairy, I was tiny, but as a kiddie I remember it being really profound. And what Lucy said really reso-nated with me, and reminded me that I do have that connection. And so I want people to see the film, and read the book, and try to cast their mind back to see if they have any of that deep-felt feeling.

HENRY DIMBLEBY: For thousands of years we have used the land to do pretty much everything – for building materials, for food, to produce energy, for clothing – you name it. We use the land to do it. And then in the last century or so, we have been digging up millions of years of stored sunlight in the form of fossil fuels. And we have been substituting that and using that to do everything. And we now need land not just to do that, we need it to sequester carbon. We also need it to restore biodiversity. And so we have to be much more strategic about: can we actually do all of those three things? And the answer is we cannot.

MIKE BERNERS-LEE: And we absolutely have to change our food produc-tion system, even just to stay still, never mind improve our yields. So change has to come and we have to adopt a dramatically more sympathetic approach to it.

DEE WOODS: Over and over and over again, indigenous peoples, African peoples, Indian and Pacific peoples, so many people saying, we have the answers … perhaps marrying some of that with some modern science, but we have the answers and I've seen it.

IAN WILKINSON: Having a regenerative agroecological system where there's massive diversity embedded in the farming system and all this land that's around us – that is surely the solution.

DW: The way forward is to go back to the land, to embrace agroecology. And those ancestral or traditional methods from pre-industrial farm-ing that were in harmony with biodiversity, with soil, and with people.

THE WAY FORWARD IS TO GO BACK TO THE LAND
AND THOSE TRADITIONAL METHODS IN HARMONY
WITH BIODIVERSITY, WITH SOIL, AND WITH PEOPLE.

Why should we care about soil?
And is this something we can fix?

HANNAH JONES: I think in life sometimes you only appreciate what you had after you've lost it. Ecosystem services have not been fully quantified, understood or valued – these are the services that support a farming systems such as pollination and decomposition – many of these services have been damaged by intensive agriculture and in so doing make farming less resilient to climate variability, and also damage the wide ecosystem. There's a recognition that we really need to respond. And there is also the additional challenge of a more erratic and variable climate. We are more reliant on the services our soils are providing us than perhaps we might have been 30 or 40 years ago, and we are going to rely on them more and more as time goes on. A healthy soil is at the foundation of a resilient farm business.

CR: Food comes from the soil. If you don't look after it, we're going to starve.

DW: Healing the soil from the violence we've done to it is really, really important. I've seen soil that was literally lifeless come back to life when agroecological methods were introduced, when villagers and people are working together to do that repair.

NICOLE MASTERS: I think what's interesting is that any tool that we have, any approach that we can adopt can be degenerative or regenerative depending on who's managing that tool. So this is why management is so important – yes, cows can be incredibly destructive and they can be one of the number one tools to regenerate landscapes. Same with ploughing. Same with the use of certain types of inputs. We can act in a very careful, thoughtful way or we can do it in a way that destroys waterways and emits greenhouse gases.

Why should we care about biological
diversity on farms? And how can we shift
away from monocultures?

HJ: There has been huge biodiversity loss in the UK from a combination of habitat destruction and the effects of landscape management. There's an ethical and moral requirement for us to look after our biodiversity, but you can also put a value on the pollinators required for land; the dung beetles, which I quite often talk about; the importance of bird species in the landscape; and all that diversity.

There is this potential conflict between where the land is used for biodiversity or whether it's being farmed. And in a number of instances, you can have a win–win, for example, with very diverse, large hedgerows. There are moral requirements – first of respect for our world, but also respect for the farmers who are providing food... There are opportunities to sit around the table and come to an agreement that some areas can't be left for wildlife, or SSSIs – corridors of land for example – but there are also cases where land can be shared between food production and wildlife habitats. It is in this area of land sharing that I think many solutions can be found.

IW: These habitats are really, really important. Most of this countryside is farmed, so everything we do here has to work alongside wildlife and not exclude it.

I've been using mixtures of different species and I've seen the benefit. I know that from a biological farming point of view, by growing different plant species together in the field, we can get this overlapping yield effect – that more species equals more yield. So, we don't have to use, necessarily, lots of oil- or gas-based products – fertilisers and things like that. It's been really interesting to see scientifically and in the field the benefits of different overlapping species. For example, we have a growing wheat and we have clover underneath it. Or where we're growing pastures we have clovers, which fix nitrogen, we have deep-rooting herbs, which bring minerals up, and we have grasses that grow at different times of the year, so we can have animals grazing outdoors for a large part of the year with really healthy forage. I recognise the yield advantages we've got with the oil- and gas-based products that we've been applying too, but the alternative way forward, and the thing we need to step on to now, is this diverse farming system – and its diversity not just in different fields, but within the field as well.

JOSIAH MELDRUM: With diversity comes resilience in ecological terms. We need that diversity in order to buffer against disease, in order to help us manage the effects of climate change – ere the only certainty is uncertainty. Farmers need to have a wide basket, as it were, of crop plants, but also of wild unintended plants growing on their farms. That diversity offers them that resilience to environmental shocks and changes, but also market variability.

241

And I think our health is also dependent on that diversity. There's lots and lots of research now and a growing interest in the gut microbiome, which suggests that we have massively simplified our diets. There are globally somewhere in the region of 300,000 to 350,000 plant species that we could be eating. But most of our plant-based protein and carbohydrates comes from the big three: wheat, corn and rice. And even within those three the number of varieties that we're eating, the genetic diversity is tiny. In any given year it's a few very uniform varieties – cultivars – from those species. That simplification of our farming system and diets does not do our health any favours either.

There's an English poet and farmer called Thomas Tusser, who wrote a poem in the 16th century about the agricultural year called *Five Hundred Points of Good Husbandry*. And it's essentially a self-help poem for yeoman farmers. See, for example, these verses from Chapter 16 (about October):

> White wheat, if ye please,
> sowe now upon pease.
> Sowe first the best,
> and then the rest.

> Who soweth in raine,
> hath weed to his paine.
> But worse shall he speed,
> that soweth ill seed.

It goes through the months of the year and tells you what you should plant and what you should harvest. He also lists things like lentils and linseed and all of these other crops that we think of as novel now. Actually, they would have been growing them back then.

Can agriculture really be part of solutions to climate change?

HJ: An estimate of the carbon stored in the soils in the UK is about 10 billion tonnes of carbon. That's equivalent to about 80 years of greenhouse gas emissions from the UK. Soils are absolutely phenomenal in the amount of carbon they can store.

We talk about carbon in the soil but it's not there in its pure form, it's integrated into decaying matter that was once living. So that could be bacteria, fungi, plants… and it's complex. It can be in sugars, protein or integrated in a chemical mix of those. If you imagine, your grass clippings are a relatively simple carbon form and that gets broken down really quickly in a compost heap. Whereas carbon in, for example, bark chippings in a path – that carbon stays there for quite a long time because it's a much more complex form of carbon to break down. And you can think of that as quite similar in the soils. There is very rapidly recycling organic matter, releasing carbon, and then there are complex forms of carbon which are resilient and will stay there for many years, sometimes decades.

IW: The idea of losing carbon in the atmosphere is crazy. I need as much carbon in the soil as I can get. That's my organic matter. That's the fertility in the soil. So when I think about some of the things that we've been doing over the years in terms of greenhouse gas emissions and reducing carbon, there are two aspects regarding climate that we can satisfy. First, is what we put on, i.e. oil- and gas-based products. We don't use them. Second, we want to sequester as much carbon into the soil as possible. So there are two things that we do on this farm that will make a difference to the climate.

And I fully appreciate that this hundred-and-something acres at FarmED is not going to change the world. But if many farmers like us are doing the same thing, collectively together we can. And you know, the farmland is about the only thing I can think of that can actually be carbon negative. So you can actually sequester carbon into the ground and no other industry can do that. So whether it's planting pastures, whether it's planting woodland hedges, we can do so much in terms of carbon sequestration and actually cool the planet.

HJ: Farmers can reduce input costs and support the environment. There is, additionally, a new emerging market, the sale of carbon credits.

FARMLAND IS ABOUT THE ONLY THING I CAN THINK OF THAT CAN ACTUALLY BE CARBON NEGATIVE. NO OTHER INDUSTRY CAN DO THAT.

So, a carbon credit is, in effect, a unit of carbon that an area of land has been able to store in addition to normal working activities. The landowner or the person who manages that land, they can potentially sell the carbon that the land will store as a credit or a unit of carbon.

IW: The climatologists tell us we haven't got decades to wait and then make decisions based on that. We need to be doing this right now. That's why we're demonstrating this farming model right here to encourage people to make changes and to support people in those changes, to get new entrants in, to find new ideas, and to use the technology.

See also Chapter 3, pages 45–61.

What's the big deal about land use?

TIM LANG: We're in a situation where we're rethinking land use, we're asking, 'What's land for?' And I think that's one of the fundamental questions I tried to address in my book, *Feeding Britain* ... It's for lots of things. It's for houses, it's for sequestration of carbon, for roads, for electricity, for water, for well-being, for food...

HD: We need to regulate and use financial incentives to make sure we use land in the right way. But in order to do that, we actually have to be honest about what land we've got... It's a difficult equation, so you have to look across the whole country at the uplands, at the peatland, at the marshlands, and say, if we were able to control this, if we were looking over the whole of the land, where would we want to be doing those different things?

MBL: Even an organic farm, you know beautifully raised cattle, might just be doing fundamentally the wrong thing with that piece of land. It might just be that the nutritional input, the nutritional output and the biodiversity gain from putting crops on the land instead of cattle might be enormous. So when we're dealing with the UK's land system, we need to consider very carefully what the best use of every piece of land is. And there are a few things we need to consider all at once. We need to look at the impact on climate change, we need to look at the impact on biodiversity, and we need to look at food production. And by the way, as we look at all those things, we need to look at how this supports the livelihoods of the people who live in those communities.

There's a sweet spot with the centre of gravity being the overlapping of three circles, one representing biodiversity, one representing climate change, one representing feeding the world population. But for each piece of land individually, there might be some pieces where you take the decision that it's so productive that this is really where nutrition should be coming from and maybe you compromise maximising biodiversity and you compromise the amounts of carbon you sequester because it's so productive. And then there'll be other land where you recognise that it won't be productive. So the best thing to do with it is to enable nature to run its course and have it teeming with biodiversity, and that will be absolutely wonderful. And there may be other land again where the balance is a bit different.

Our land is such a precious resource. We don't have that much of it in the UK, but it's so important to us for all the things it can deliver. And it's complicated to know what to do with it. We should put proper resources into working out what the very best use of all that land is. There's a whole load of emerging research that we so critically need: what can or can't be achieved through carbon sequestration, through livestock; how do you grow crops in a productive way without so much use of fertiliser and pesticide and so on? All these questions are just not understood as well as they need to be.

HD: The government announced that it would produce a land-use framework, which is just fantastic. It could be transformational. Let me give you an example. A farmer who farms on rich peatland in East Anglia who's getting mixed messages. The rich peatland produces a lot of food, but the farmer's told it's releasing a lot of carbon from the peat. And that the biodiversity is not good. How can a farmer negotiate these three things? The answer is we need a land-use framework to do that.

See also Interlude I, pages 40–43

How do we reconnect to the land?

DW: It starts with who owns the land, who has access to land, and for people like myself it's difficult even if you have money.

IW: One of the biggest reasons people can't get into farming is no land.

DW: Just sitting here and looking at the landscape, what I see is the history of enclosures and private ownership, because that's what all these hedgerows represent.

Land value is at a premium. For example, the land Granville Community Kitchen sits on is probably worth about £20 million. There's no way a community could buy that. Do we need land taxes? Do we need reform in terms of having landowners give some of their land away? There is a major barrier for young people and for young people from diverse backgrounds to access land. There are so many people who want to farm – it's one thing growing food in a community garden, but to actually farm, they can't.

I was talking to the local councillors about opening up land, because as I was walking I could see there were derelict plots. Who owns that land? There's lots of conversation to be had with local authorities, with landowners … because no one landowner can really farm all of it.

SATISH KUMAR: I believe that we need more people on the land. At the moment in England, hardly 2% of people are on the land. We need maybe 20% of people on the land. Land should be a source of good employment and we should not have this factory farming culture. We should have proper agriculture. We should not have agri*business*; we should have agri*culture*.

DW: I think of the Landworkers' Alliance, where I'm a director and Food Justice Policy Co-ordinator. Things have shifted within the Alliance – I would say at least 13% of our members are black or people of colour. So, you know, I think things are changing, but there needs to be so much more work to remove some of the barriers. I think the main one is about being in rural spaces and how unwelcoming it is, and unfriendly, and it can even be violent. So there needs to be some work done in rural communities and the wider farming communities to break down those barriers of racism.

WE SHOULD NOT HAVE AGRIBUSINESS; WE SHOULD HAVE AGRICULTURE

CM: Yes, absolutely. When making the film we did try to seek out diversity where we could, and we're really happy to be able to show more women in the industry, but the fact that we didn't find any farmers from diverse backgrounds to work with is perhaps a comment in itself.

IW: What we need is new people, new entrants into the farming system. You know, I'm 50, I'm the average farmer and I need new energy … I need to work with young people. I need new entrants, and they don't have to be young all of them. They can be a range of people. It can be retraining. Nothing wrong with any of that. But I just think we need to have more people on the land, not fewer.

You know, robotics is great, AI is great, big data's great. But we still need people on the ground literally connected to it to see what's happening on a daily basis.

CR: There is a place for technology, where it's augmenting what's already done by humans and not replacing them.

JOHN PAWSEY: I think we're going to be using technology more to solve some of our problems, but we're going to have a much more intimate relationship with the minutiae of soils and of plants and of all those agronomic aspects that we've probably not been engaging with enough because we've been too busy out there, you know, sitting on tractors and being divorced from it.

DW: So for me, it's about building equity and if you start peeling back the layers of this rather convoluted food system that we now have, you will see it is about issues of power and disconnection and loss. And part of my work is about helping people to reconnect to land and reconnect to knowledge. And particularly with majority-world people who, because of colonialism, have had that knowledge completely interrupted and now feel loss.

JM: Because we industrialised really early and we urbanised really early, we lost that connection with land and landscape and sense of place. And it means our diets have become industrialised and simplified in a way that really hasn't happened to the same extent in much of the rest of the world. So even in other parts of Western Europe, like France and Italy and in Germany, people often still have a relative, probably, who's got a smallholding or a farm, and they've still got that kind of connection to the food culture. We've almost completely lost that. We're urbanised into the high 80% of our population and even through COVID, where some people are moving back out to the countryside, overall there was still net urbanisation going on.

The simplification and the industrialisation of our diet is really damaging. So, for example, where in Italy they might celebrate mushrooms or in France chestnuts, and they'll have festivals, and you might see them on the telly … here we celebrate the biscuits that Victorians invented. We celebrate the industrialisation of our food system in a way that most other countries just don't. And so the diseases of chronic ill health that come through diet in the UK are particularly obvious and we have the lowest quality of life expectancy in Europe. And the same goes for the United States where they've also industrialised their diets much as we have.

SK: We started to think that working on the land is something backward. If you want development, progress, being educated, being smart, being clever, you don't work on the land. You work in the factories or industry, or you become a lawyer, or you become an accountant, or you become a civil servant. So the English have unfortunately, or British, I should say, but English especially, have seen work on the land as something for the uneducated. And that has been an unfortunate situation.

I see soil as a miracle. Six inches of soil feeds 8 billion people, not only 8 billion people, but the rest of life. All the animals, bugs, insects and even trees are fed by the six inches of soil. It's a miracle. And yet we think soil is dirt, and dirt is dirty. Dirt is not dirty. Dirt is the source of life. And that's my passion.

Nature and humanity are not separate; what we do to nature, we do to ourselves. If we pollute water, we have to drink it. If we pollute air, we have to breathe it. We pollute our soil, we have to eat food from that soil. So nature and humans are not separate.

DW: I think this is what is missing here. No one really envisions a future where we can live together as people in harmony with ecology. Somehow, we're missing that.

NATURE AND HUMANITY ARE NOT SEPARATE;
WHAT WE DO TO NATURE, WE DO TO OURSELVES.

How do we regenerate communities and relationships? And can we build relational supply chains?

IW: We all know we need to change. We're not quite sure exactly how that change should happen, and we know there are barriers, but understanding each other's perspectives is the starting point.

DW: We have to understand the interconnections; we damage the soil, then that impacts us, that impacts the biodiversity. We need to understand that we need to be healing all those relations and not just focusing on soil. We're regenerating soil, we're regenerating people, we're regenerating relationships, and we're regenerating values.

IW: Whether you're thinking of food production or mental health benefits, these farms are a massive resource. My role at FarmED is to bring people together to make connections between different actors in our food system.

    We built a farming centre and we've got great facilities now to welcome people here. And well, we get all sorts of people. I would never have expected so much diversity in the people when we started, which is great. We are visited by primary schools, by universities, academics, all sorts from around the world and we also get politicians, decision makers, civil servants, farmers and growers of course, and it's great that they mix with each other when they come. You know, by sharing ideas and kicking things around, we often make a lot of progress very quickly. And it's all well and good that we have formal lectures here, we have lots of courses, which is great, but it's often the conversations over a cup of tea and a bit of cake that really count.

    One of the early entrants here was a community supported agricultural scheme (see Chapter 5, from page 127). A bunch of supporters pledge money every month to the growers. It's great for the growers who can make a living wage, they don't have to worry about the markets, and the supporters get to know the growers, how they grow the food and they pick it up every Friday. It's amazing. And for me, as a farm, that brings everybody down the farm drive on a Friday afternoon. The micro-dairy can sell its milk. The honey from the apiary can be sold. The bread from the local bread maker can be sold. It's like a circular economy the whole time on the farm. And that's really exciting.

DW: It is extremely important that people are able to access fresh food and that is what we do at Granville Community Kitchen. We

249

provide a space of safety, belonging and dignity where people can access cooked meals. We do some food aid for people who cannot access work or welfare because of current legislation, so that includes migrants and a lot of asylum seekers, and we grow food. We grow food that goes into our kitchen, but also into our veg boxes, and that was one way we found that we can enable low-income groups to access organic, culturally appropriate, fresh food.

It is very slow work. I've been involved in this south Kilburn community for well over 30 years. And we started small. Even though we had our ideas … all our strategies, how we work up our meals, our gardens, everything is co-created with the community and the community takes an active role in delivering things. We're not looking for ownership. We seek to steward the land in terms of biodiversity, in terms of health for everyone, in terms of creating jobs. So that's what Granville Community Kitchen have done.

You know, we always commune around food. For me, food is that one thing we all have in common, and it can break through any barrier. So it doesn't matter if we don't speak the same language. Food has a language of its own.

VH: Farmers are not being paid enough, it's absolutely clear. They get less than 1% of any profits (Figure 8.1) and overall only about 10% of the gross value added of the food system. Often they get too little to make a decent living and the profits are just sucked out of the system beyond the farm gate.[3] And that's the retailers, the manufacturers, the intermediaries and the retailers are taking all the money that we pay as consumers. We need to change the supply chain and we need to change how workers are treated.

JM: We tend to think in a very linear way. We've got the farmer at one end of the chain, we've got the consumer at the other end and, in the middle, we've got processors or brands or supermarkets and there are a lot of barriers between the two – very often intentionally. If we think more in a networked way where farmers are at one node,

FOOD IS THAT ONE THING WE ALL HAVE IN COMMON, AND IT CAN BREAK THROUGH ANY BARRIER.

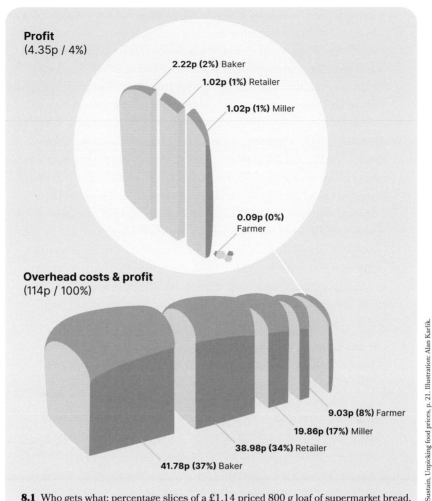

Profit
(4.35p / 4%)

2.22p (2%) Baker

1.02p (1%) Retailer

1.02p (1%) Miller

0.09p (0%)
Farmer

Overhead costs & profit
(114p / 100%)

9.03p (8%) Farmer

19.86p (17%) Miller

38.98p (34%) Retailer

41.78p (37%) Baker

Sustain, Unpicking food prices, p. 21. Illustration: Alan Karlik.

8.1 Who gets what: percentage slices of a £1.14 priced 800 g loaf of supermarket bread.

and we've got consumers who might also be farmers, and we've got researchers, and we've got retailers, and we've got wholesalers like ourselves [Hodmedod] that are all connected and have a free flow of information. Then there's a much richer experience for everyone involved and we can understand the challenges, but also see opportunities that currently don't seem feasible.

TL: The task system that's been built up over the last 40 years was borrowed from the car industry, which is a system of incredible efficiency. Why have a warehouse keeping lots of tyres, motors, gearboxes, steering wheels, windows, if you can have them arrive at the factory where you put them together to make a car just in time. And, basically, Tesco followed that. And Tesco doubled its market share very fast when it started applying this and it's never lost that leadership. But the system that was pioneered here in Britain is now spread across Europe and elsewhere in the world throughout the food system.

The retailer is king, the retailer is sovereign. When people talk about food sovereignty, I laugh because actually it's not about bottom-up groups. That's a great argument that came from the developing world and peasants' movements, but it just does not apply in the rich world. The food sovereignty in Britain is retail power. Nine retailers have 94.5% of the market, and COVID expanded that because basically the government allowed the hospitality sector to be destroyed.

A lot of energy has been put in by the critics of the food system to try to create what has been called alternative food networks, farmer's markets, box schemes, etc. I mean, they're magnificent, I love them. I was a trustee of Borough Market in London for 10 years, which is properly wonderful. It's terrific, but it's tiny.

VH: The key thing is to make sure that you have strong regulation of supply chains so that you have legally binding codes of practice. So all of the supply chain has to pay when it says it's going to pay. It pays decent amounts for the produce it's buying and it doesn't force on farmers and growers ridiculous cosmetic standards – all these kind of things need to happen. We also need to be building alternative routes to market for farmers so they can find alternative places to sell their produce. And to sell the more diverse produce that regenerative, agroecological farmers generate. So building the alternative and regulating the current supply chains.

We need to be building regional food systems with all the infrastructure that will allow farmers of all types, the commodity producers, the horticulturists, the meat producers, to actually get their produce to market within a region. So you've got that diversity. It's not going huge distances, it's not overprocessed, and it's giving them a fair wage.

JM: We began as a community project in Norwich which asked: Can the city feed itself? We realised that it could and we set up some demonstration projects, growing vegetables and milling grain in the city. But we also knew that this new sustainable rotation needed to include a lot more legumes – both in fertility building leys and in harvestable pluses – beans and peas. We put that off that work to the end because it felt the most difficult bit of the project, but we later realised it was actually the most exciting part of the project. How can we transform arable rotations to include more of these nitrogen-fixing, carbon storing, very healthy and nutritious crops? We did some trials. We took fava beans that would otherwise have been exported to Egypt and other places and discovered that actually people in Norwich were keen to eat them.

Fava is a crop that has been grown here for 2,000 years, but it fell out of favour as a food 400 years ago. Instead we feed it to animals and export a little. In 2012, having done that trial work and having tried to place the beans with wholesalers who weren't very interested, we realised that we needed to set up the business ourselves. Beginning with beans, and then thinking about how we could create a more complex agroecological rotation around those pulse crops.

We're a very small business, if a farmer were to phone up tomorrow and say 'We've begun this regenerative process', it could take a couple of years before we're in a position to buy anything. But the conversation is ongoing. We're building that trust relationship and that eventually we'll say, yeah, let's do that, we're ready and you're ready. The way that we work with our farmers is we have a conversation at the beginning of the year about what they want to grow and how much of it they'd like to grow, and we agree on an outline specification for how it should look in terms of quality.

And then at the end of the year we'll have another conversation – how did it actually go? We'd have that conversation with John Pawsey (see Chapter 4) about the lentils that he's grown for us. What did they yield? What did we have to remove from them? And what is a fair price that reflects his cost of production and incentivises him to grow lentils next year, because we don't want him to just walk away and say: well, that was fun, but I'm never doing it again. We want to ensure that we can have an ongoing relationship with John, but at the same time we have to work out what our customers are prepared to pay for those lentils. And there's a midway between those two routes. And that's the only way, really, that we can work out how to pay a fair price and sometimes we just have to acknowledge that a particular crop on a particular farm just isn't going to work in the long term.

253

We identify the farmers on the packaging and celebrate the differences between the crops on different farms. So if two farms are growing the same variety we will still identify them as different crops and they will still taste subtly different, which always surprises people.

We learned a lot from the first farm we worked with, Mark and Liz Lea at Green Acres in Shropshire. They'd been growing peas for animal feed and wanted to grow them for human consumption. And we were looking for a farmer to do that. So he was our first pea grower. He'd had a lot of experience marketing meat locally, through a meat box scheme, so he was used to getting feedback on those products from the farm but had never had anything from a field crop. We put his name on the packet and then he would see what our customers were doing with it on social media and it was amazing for that kind of connection to happen.

And then we won a prize – the Soil Association Boom Awards, the best store cupboard item – for his peas. He was asleep somewhere in Shropshire while we were down in London celebrating his peas, but we tagged him all of the posts about this amazing thing that happened. And so he woke up in the morning, and he said he looked at his phone and thought the barn must have burnt down because he had so many notifications. Then he looked and he realised it was his peas, and he said it was part of this process of him feeling that those acres that were growing food again and they weren't a commodity. And he had that connection. And it really changed his relationship with the farm.

Now he grows 14 or 15 varieties of wheat in any given year in strips (growing different crops in sections of the same field). He's bought a mill. He sells it locally. It's completely changed his relationship with arable cropping.

Another massively important cog is the shop. Shops can have a really positive role in the community. It's a place where people meet and talk to each other. It's a way of connecting land to the consumer without taking up too much of anyone's time. In the last 50

WE'VE GOT REGENERATIVE FARMING. WHAT ABOUT THE REGENERATIVE RETAILER THAT PLAYS AN ACTIVE ROLE IN REGENERATING OUR SOCIETY OR COMMUNITIES.

years we've had the supermarkets which have taken that to a logical extreme where they have just become these big sheds where you don't even necessarily talk to anyone at the checkout because there isn't anyone, there's just a machine waiting to scan your products. And I think what we need is a return to the regenerative retailer. We've got regenerative farming. What about the regenerative retailer that plays an active role in regenerating our society or communities – connecting us with food and farming and bringing that vitality back?

What should we be eating?

MBL: So when you're looking at food, of course climate is only one of the environmental components that we need to be thinking about. But just looking at the carbon for a moment, if you take the average UK person's carbon footprint, it can be divided up into roughly four, almost equal quarters. And one of those quarters is food.

It's hard to get a good enough understanding of supply chains. The food producers and the supermarkets could help. They could do a better job of giving us meaningful, accurate, non-greenwashing information about what's been going on in the production of food. And we as consumers can take responsibility by trying to learn about it. We can't always get instant knowledge on everything, but we can try to build it up. We can learn about one food after another, maybe occasionally look up a company whose products we buy and see what people are saying about them.

The supermarkets are very skilled at nudging the probability that we'll choose this product over another. If one product has got twice the margin of another product, they know how to make it twice as likely you'll choose that one. And they could exert that influence to make us more likely to choose the plant-based and more sustainable options.

HJ: I think understanding the stories of the producers, the connectivity between the buyers and the producer, is essential. One of the major challenges is making it profitable on a relatively small farm. And this is where customers can make a real difference – in recognising the great work that is being done on a number of these farms.

IW: We need to have a lot more understanding of how farming is done so that when you and I go to the supermarket, or wherever we get our food from, we can pick and select from a farming system that is truly sustainable, not just one that says it is. Let's find a language that we can all understand. Find systems of labelling that we all understand.

JM: The livestock question is a big one. Regenerative systems don't produce as much meat, they just don't. And they won't. The role of livestock is nutrient cycling. There is a lot of talk about how we can produce more meat, but we can't. Many campaign groups are now coming around to the perspective that we will have to eat less meat. And I think one of the things that the regenerative movement perhaps hasn't properly come to terms with yet is just how little animal products these systems can actually produce. So, as an example, in the UK we eat 1 billion chickens a year. It's huge and there's no way we can produce that through regenerative systems. You know, it's completely dependent on grain-fed poultry in sheds either in the UK or in other parts of the world. We'd have to massively reduce our chicken consumption, probably back to our postwar years when chickens were really valuable.

And we've got a dysfunctional dietary relationship with meat which sees us devalue those sentient beings that we share the world with. We need to change that relationship and our diets need to reflect that. That's a very difficult conversation to have with the public. It's very difficult to have that conversation with the farming community as well because farming is not just a job, it's a vocation, it's an identity. If you're part of a community that produces particular crops or particular livestock, a challenge to that business model is really significant.

MBL: In the meantime, there are some simple rules which are: less meat and dairy, and if you're going to have meat and dairy organic is probably a good rule of thumb for how it should be produced. Make sure your food has not travelled on an airplane, which is a little bit difficult to tell whether it has or not, but if it's come from a very long way away and it's something fragile like a strawberry or a raspberry or something

WE'VE GOT A DYSFUNCTIONAL DIETARY RELATIONSHIP WITH MEAT WHICH SEES US DEVALUE THOSE SENTIENT BEINGS THAT WE SHARE THE WORLD WITH. WE NEED TO CHANGE THAT RELATIONSHIP AND OUR DIETS NEED TO REFLECT THAT.

that won't last three weeks on a boat then it will have probably been on an plane. And we can look at the packaging and ask ourselves whether this looks like a sustainable way of storing food or not. We can begin to ask those questions. But I think a really good start point is just when you go around the shop, just start asking yourself the questions, just start being curious about it. And even if you can't answer them then at least you've started looking for the answers.

How do we make policy to support sustainable food and farming?

TL: The history of Britain is a history of a country that decided in 1846 not to feed itself, but to use its colonies to feed it. And we're still in that legacy of politics. One of the things that I argued in my *Feeding Britain* book was a need to nail down what we mean by self-sufficiency and food security (see also Interlude III). Being self-sufficient is not necessarily a good thing. I can't imagine Britain doing without food that is spicy, but we don't grow turmeric or cayenne pepper and so on. And there's a pretence in Westminster that we're separate from Europe. No, we're not. Not at all. So where's the food going to come from? What's the long term? Because we're now responsible for our own food security.

So for the last 30 years, Britain has operated its food economy within a single market. We're now outside that single market, but yet we're still being fed by it. We've got delays at borders, we're paying taxes at borders, and for exporters from Britain it's become very difficult indeed. And more attention now is coming back, I think rightly, to the food trade gap. It's currently £26 billion in the negative. If it wasn't for sales of whisky and biscuits, frankly, it would be far higher, because those are the only commodities in which we're in surplus … we import more meat than we export.

VH: I've been working on food and farming policy for 30 years and we absolutely know what needs to happen to solve the climate crisis, the nature crisis and our obesity crisis. What we need now is urgent action from the government to provide support for farmers to do nature-friendly farming – go organic, go regenerative. But they need the support to do that because it costs a lot. There are big risks involved, so they need that support. They need the government also to build up alternative routes to market and they need the

government to regulate the supply chain, so they pay. These things we know absolutely need to happen. Let's get on with it.

What we need to get this off the ground is for new entrants to be able to do what they need to do. We need the proper training and demonstration and advice. We need affordable land and we need finance so they can actually get on with it and have adequate resources. And they need access to markets as well.

HD: I am the author of the National Food Strategy, which was an independent report for the government, setting out how we could create a food system that fed us all nutritious food, at a reasonable price, without destroying our health and destroying the planet. So we recommended four pillars; four things that we need to do.

One was to escape the junk food cycle. To put in place policies that broke that commercial link. The second was to support people who are in poverty, to support their diets. You need to raise them out of poverty, but while they are there, support their diets. The third was to make the best use of land. The land is not there just to produce food. It's got to restore biodiversity, it's got to sequester carbon, it's there for recreation. And we need to be much more strategic about how we apportion land to different uses. And then, finally, we need to change our food culture because there's only so much you can do through production and you need to change how people consume food as well.

And as a result, the government put into place holiday activities and food programmes. So now any child on free school meals can access both food and good activities throughout the holidays. Fantastic improvement. They improved the money spent on Healthy Start, which is a scheme that gets fresh fruit and veg to families with young children who are struggling. The levelling-up white paper put in place some of my recommendations on education and on trialling things within communities to change the food environment.

MBL: Dietary change is an opportunity to bring about a healthier population and a fundamentally more efficient food production system that's good for the economy.

HD: Government responds to things with different government departments putting in place separate policies. So it is very difficult to actually create a holistic, meaningful strategy response from governments. The nature of policy is it's always tactical. It's difficult to do quite bold things. You need a government that has a lot of political capital, that has the ability to be brave and bold.

MBL: There is a really important place for government intervention. The government already intervenes into the food system in a huge way through subsidies and there's opportunities there to change the way those subsidies work, so that they do a really good job of incentivising the best agricultural practices. So that's a huge and fairly simple step to take.

IW: The Environmental Land Management scheme (ELMs) will be helpful to some farmers. It will enable a transition from a simple system to a more diverse system, and it will pay farmers to have cover crops and things like that, which is helpful. And it will, on a landscape scale, also provide money to farmers to collectively work together in wider area, a river valley for example, to deliver wildlife benefits. So that's really useful.

TL: For me, the critical issue is why are we even subsidising farming? Surely farming should be getting the true value of what it produces. We're playing around with subsidies and arguing about subsidies when we should be saying pay the farmers to grow decent food to feed us.

WHY ARE WE EVEN SUBSIDISING FARMING?
WE SHOULD BE SAYING PAY THE FARMERS TO
GROW DECENT FOOD TO FEED US.

We need a bit of imagination to address this crisis because the political imagination has got to change from assuming that charity is the answer to food insecurity. It's not, and it never has been. When it comes to alternative food networks, they are very small. They're inspiring. They show us that other ways are possible. But until you deal with the corporate power of nine retailers who have 94.5% of food retail sales in Britain, you're not going to have a level playing field. The politicians think food should just be left to Tesco. That's the hard reality. Until we break that consensus in the policymakers, we're going to get nowhere.

If I have one minute with the prime minister, I'll just say one word: horticulture. We have to really invest and skill the population, not just a few horticulturalists. We've got to skill people back to being able to grow food that they can eat.

DW: In this country we do need to scale up. We do need to be producing more food, especially within the horticultural sector. But we need to be doing it in a considered way and an agroecological way so that it becomes the norm in this country once again. It's not impossible. What we do need is commitment and action from our governments. What we do need is finance for the agroecological movement and transition. If we really want to have that major change, we need that investment.

Instead what we have is protracted inaction by our government to deal with poverty. We need to make good food accessible to people, affordable to people, but there needs to be work done around campaigning for better wages and a better welfare system.

JM: Very much so. The question is always framed wrongly. It's always framed as: how can we make food cheap? Whereas the question should really be, how can we ensure that everyone can afford good food? And good food is food that's produced without compromising people's capacity to eat next week, or the week after, or in future generations. And as a society, as a democracy, we need to arrive at a level of governance that enables everyone to access good food.

How does this fit with wider social change?

TL: Do you want food to be a key element in how we live and how we live well? Or do you want food just to be fuel and cheap?

DW: I think there needs to be a wider systemic change around wealth in this country. We need a new social contract, basically, that ensures that there's a more even distribution of wealth, but within our food system as well, we need a new contract with our ecology with each other to ensure that everyone can access food.

TL: I don't think regenerative agriculture has got the answers to everything because I think the answers are not actually about the land, the answer is about society. Regenerative agriculture has its place as being part of what we must do, but I don't see that happening at scale sufficiently unless the people are involved. I'm a social scientist, that's my bias, but I was a farmer and I'm very keen on really good ecological management of the land. So I understand the arguments for regeneration. I was an organic farmer. I'm president of Garden Organic. I'm deeply committed to good, ecologically based land management. But the crisis is people, the crisis is how we run an economy.

DW: The corporations and people with big money, the philanthro-capitalists, have the power, so they are setting the agenda. So just as we have that groundswell of people recognising that governments will not do anything in the immediate for us, which is why we've taken it upon ourselves to create the systems and the communities and the infrastructure that we need at local level, we also need to be building that true democracy. I don't know how it's going to come about, but

WE NEED A NEW SOCIAL CONTRACT, A MORE EVEN DISTRIBUTION OF WEALTH, WITHIN OUR FOOD SYSTEM WE NEED TO ENSURE THAT EVERYONE CAN ACCESS FOOD.

that's what we need to be working towards. And that to me means having those conversations with businesses; having those conversations with philanthro-capitalists; having those conversations with big landowners, because the existential crisis that we're in is going to impact us all.

JM: I heard the head of environment for one of the big supermarkets at Groundswell explaining that they couldn't really change things because Tesco only controlled 28% of the retail market – only 28%! So they couldn't really take action on their own in order to make changes happen. What she really meant was we have a shareholder model. That means it's impossible for us to change because anything that affects our profitability is detrimental to the shareholders and, therefore, we can't do it. And that is the big problem.

MBL: If you look at history, we're capable of actually very great levels of empathy and cooperation with each other. What we haven't yet learned to do is scale it up to the global level, and that is the absolute pressing challenge for humanity now, we need to learn to cooperate properly at the global level.

Neoliberalism is based on the idea that the way forward is, fundamentally, by playing to human greed. And that is absolutely poisonous, it's the opposite of what we need to do. And we need to challenge that, reinvent it and get out of it. And I hope we can do it without having to completely melt down first.

Under the neoliberal system, nothing will work. But the idea behind ESG [environmental, social, and corporate governance] is it's supposedly a measure of the extent to which a company is looking after people and planet and, if you could do a really good job of that analysis, then it would be worth something to the investers and bring trillions of dollars pushing for a better world. The problem is that it'll only be enough if it's also backed up by a culture in the investment community of wanting to do the right thing.

NEOLIBERALISM IS ABSOLUTELY POISONOUS,
IT'S THE OPPOSITE OF WHAT WE NEED TO DO.

SK: So in order to make a regenerative economy we have to create a cyclical economy, a circular economy. Whatever you take from nature, put it back in nature, and it will become soil again. So, for example, you have bananas. Bananas have leaves or skin. Banana skins should go back into the soil. You have oranges, orange peels should go back to the soil. Our industrial system is extractive. It exploits nature and it destroys nature. Tremendous amounts of waste, tremendous amounts of pollution. If you have a circular economy, there'll be zero waste, zero pollution, an abundance of food and an abundance of everything else.

What in our language we call economy is not really economy, we are misusing the word. What we call economy is actually money.

Ecology means knowledge of the household. 'Eco' in Greek means household and, in the wisdom of Greek philosophers, the entire planet is our home. We have knowledge of household, 'ecology', and management of the household, 'economy'. So I said to the London School of Economics that you are teaching economy, how to manage our household, without teaching ecology, knowledge of household – how are you going to manage something which you don't know?

We have come to believe that money is more important than soil, and money is made by industry, by manufactured goods. So you produce more cars, more computers, more cameras, more airplanes and you build railways and you build motorways and you build airports. And that's where the money is made. Food and farming is for subsistence, for living, not for money so much.

In Indian culture, and also all the traditional indigenous cultures around the world, soil is seen not as a resource for the economy but as a source of life itself. If you see soil as sacred and soil as the source of life, then you have more reverence for soil. I was brought up in India. My mother was a farmer and therefore I learned to respect the soil, the view of the soil, and see soil as a source of life and not as a debt or as something for poor people to work on.

The way to bring respect for the soil again is to bring the education of soil, an education of nature, in our schools and in our universities. Every school should have a garden. Every school should be associated the farm. Every university should have a farm. And I would say that even parliamentarians, the members of House of Commons and House of Lords, should have a garden and they should work on the land. I was delighted to be invited by King Charles, when he was Prince Charles, to Highgrove. He gardens and he loves gardening. That was wonderful.

I think we need to bring dignity back into agriculture. Our young people don't know the names of trees. They don't know the names of birds. They don't know the names of butterflies. They don't know anything about nature or agriculture or food. How are they going to bring that dignity to it for farming? So our job is to have a campaign and movement to bring soil back into education.

Agriculture should be for everybody. All those who eat must participate in growing food.

What gives you hope for the future?

MBL: If you want to feel empowered, if we want to have some agency on this problem, the question for us to ask as individuals is: What can I do to help create the conditions under which the world can undergo the big systemic change that we so urgently need? That's quite an empowering question. There are the obvious things you can do about living more sustainably and reducing your carbon footprint and so on, but there's much more you can do as well. You should ask: How else can I exert my influence? How do I influence my workplace? How do I vote? How do I influence my MP?

VH: What gives me hope now is this fantastic ecosystem of people doing amazing work, like the soil under the ground, working in complex ways to deliver food and nature. You've got people aboveground working together, the farmers working with the communities, working with local people. I think those kinds of movements are really what give me hope.

CM: As this book was going to press, we were contacted by someone in Poland who is keen to see the film as she feels her neighbours don't understand ecological farming and she wants to show the film to generate a positive discussion. This is exactly the reach and impact we hoped the film would have. We have had screening requests from groups of all sizes: a knot of friends, villages, towns and large organisations. People want to discuss possible ecological solutions and see the film as a vehicle to enable them to do this.

We have passionately created *Six Inches of Soil* with a very supportive community and now this is growing. The time is right for a film and book that contain realistic solutions, don't over promise, tell a human story and connect the audience with soil, farming and nature. We all believe in nurturing our soil, farming, nature and communities, and that a long-term vision needs to be adopted that will build a resilient and diverse food and farming system. Regenerative farming and agroecology needs to be centre stage of policy, it has the answers and it deals with the huge amounts of waste that is created in the current system.

We have been asked what is the one thing you would tell people to do having seen the film, or read this book. I can't keep it to one, so here's a few:

- talk to your local farmer, get to know who grows your food
- touch and smell soil
- go outside more and take time to stop and observe
- work with communities from farm clusters to local eco groups.

THE PEOPLE WHO'VE BEEN DOING THIS WORK SO FAR HAVE BEEN DOING IT MOSTLY UNACKNOWLEDGED WITH VERY LITTLE SUPPORT. BUT THEY HAVE THE ANSWERS AND THEY HAVE THEM READY. WE JUST NEED TO SUPPORT THEM.
~ADRIENNE GORDON

Epilogue – Act now

Join the growing movement **www.sixinchesofsoil.org**

Share
By talking to others you'll find common cause and encouragement, and who knows where that will lead.

- Talk about the film and book with your family, friends, fellow farmers, work colleagues, everyone.
- Post on social media.
- Share this book. Lend or give it to a friend. Sell it second hand. Donate to a book swap. Buy and gift extra copies.

Watch
There are film screenings running throughout 2024. Check the website to find one near you.

- Already seen the film? Excellent. Go again and take someone who hasn't.
- Not seen the film? What are you waiting for?
- Are you part of a group that would be interested, or know of a suitable local venue? Host a screening.

Buy better
Use your purchasing power to drive change.

- Buy less meat. Avoid industrially produced meat. Buy better quality meat.
- Buy more fruit, vegetables and pulses.
- Buy locally produced, buy organic, buy from regenerative sources.

Change
Be bold, make a change, take the first step. If you're already on your journey, share what you're doing to encourage others.

- Farm or manage land? Start using regenerative methods.
- Manage a company? Examine your supply chain. Cut out environmentally damaging suppliers and those that finance them.
- Work in finance and policy? Facilitate change.

Lobby
It's not just about being a good consumer, become an active food/farming citizen.

- Join a regenerative farming organisation (e.g. BASE, PLFA).
- Support by becoming a non-farmer member (e.g. NFFN, Landworkers' Alliance).
- Drive soil health, good food and regenerative farming higher up political agenda.

Grow
Get your hands in the soil.
Connect with your community.

- Grow and eat something you've produced.
- Make compost at home or as a community (explore methods such as bokashi/vermicompost).
- Set up a gardening club at your local school or in your community.

Discover
There's a wealth of resources and advice out there to inspire and guide.

- Sign up to the Six Inches of Soil and 5m Books mailing lists.
- Dig deeper on the resources page https://www.sixinchesofsoil. org/impact-campaign, into the Further reading and check out the Partners in Change (Appendix I).
- Find your local regenerative farm (https://regenerativefarmersofuk.com/map/), your local pasture-fed meat supplier (https://www.pastureforlife.org), your local CSA (https://communitysupportedagriculture.org.uk), your local food suppliers (https://www.bigbarn.co.uk) and Better Food Traders (https://betterfoodtraders.org/find-a-better-food-trader/).

APPENDIX

Partners in change

We have partnered with a number of farming and food organisations that reflect the mission and values of Six Inches of Soil. Our partners are guiding development, creating access and supporting the social impact of the project.

BASE-UK

The Country Trust

FarmED

Food, Farming & Countryside Commission

The Gaia Foundation

Groundswell

Nature Friendly Farming Network

Pasture for Life Association

Organic Farmers & Growers

The Soil Association

Sustain

Sustainable Soils Alliance

BASE-UK was founded in 2012 and is a farmer-led knowledge exchange organisation for individuals interested in regenerative agriculture and passionate about the sustainability, health, and growth of our soil and therefore our industry.

BASE stands for Biodiversity – Agriculture – Soil – Environment and the UK group was formed to run in parallel with the innovative and influential group started in Brittany by Frederic Thomas in 2000.

The group follows the principles which are fundamentally about carbon management and health in soil based on three core principles:

- Minimum soil disturbance.
- Residue covers on the soil.
- Rotations.

Our membership consists of no less than 80% farmers so that we can provide a means to share the wealth of experience, knowledge and collective learning of our members and we encourage members to present their experiences and host farm visits.

We are wholly independent from any business or organisation and are self-funding; this is much appreciated by our members.

We were contacted by the Six Inches of Soil team in the very early days of the film's creation for contacts for the film, and sponsorship. Some of our members including Andrew and Anna Jackson, Tom Pearson and John Pawsey feature in the film along with others such as David White who provided support and technical inspiration behind the scenes. We are delighted to have been able to contribute towards this inspiring project. To find out more about BASE-UK, visit www.base-uk.co.uk.

Left to right: Andrew Jackson, Anna Jackson, Tom Pearson, John Pawsey and David White.

Country Trust

Connecting children with the land that sustains us all

www.countrytrust.org.uk Registered charity: 1122103

Most of us are unaware that we are totally dependent on the land for our survival. We are oblivious of how the food we choose to eat impacts on the world around us. The natural systems that provide our food, water, climate and clean air are in crisis as a result of this widespread and multi-generational disconnect. Consequently, we as humans are failing to thrive, with the most disadvantaged in society most affected.

Meanwhile, in the national curriculum there is no consistent provision for every child to explore and discover first-hand the connection between their actions, their health and the health of the planet. We are not equipping our children to grow to become custodians of a thriving planet.

Since 1978, national charity The Country Trust, has worked to change this, connecting disadvantaged children from the earliest age with the land. The charity's expert educational practitioners bring together primary teachers and farmers to empower children to be confident, curious and create change. Tens of thousands of children each year experience sensory, hands-on learning experiences including Farm Discovery day visits, year-long Food Discovery programmes, Countryside Discovery Residentials, Farm in a Box and Plant Your Pants, a national soil health campaign.

The Country Trust fundraises to support children whose potential is only limited by their access to opportunities. Poverty limits access to good food grown well, green space, and impacts health, attainment, social mobility and lifespan. In 2022/23 The Country Trust connected 70,192 children with the land across 1211 schools in areas of high disadvantage.

> I loved the time when we grew stuff in school because I'd never done it before. Our garden has all pavements, so we didn't grow anything. I have a disability sometimes that makes me use a wheelchair. It annoys me and gets me down but when I was growing food, it made me feel stronger, like in my mind and my body. I loved it. We have some pots that we grow a bit of salad and some herbs in now, but I'd love to grow more stuff because I like how it makes me feel. Maybe I'll have my own garden one day to grow my own food that I like to eat.' (Child P – Year 5)

Over the last 75 years, farmers have responded to policy by using a combination of fertilisers, pesticides, plant and animal breeding along with increased mechanisation to double the quantity of food produced from farmland. As impressive as this has been, there have been unintended consequences that have arisen. So, the question is, how do we farm in a way that produces enough food and doesn't destroy the world?

The millions of people who eat food have become detached from the few thousands of farmers who grow it for them. To reconnect with farmers and help support them to rebuild soils and to nurture healthy food systems, we need a place to reconnect. FarmED doesn't have all the answers but it provides a place where ideas are shared, not judged and where connections are made between the different actors through the totality of our food system. It is optimistic, inclusive and welcoming to all.

Since 2013, FarmED has created an award-winning demonstration farm, showing the contrasts between conventional, regenerative and agroecological farming systems. Visitors are shown fertility-building crop rotations that increase soil health whilst still producing nutritious food. Agroforestry, pasture-fed livestock and a natural flood management scheme are integral to the demonstration showing visitors the potential for climate resilience. There is a wide variety of nature in meadows surrounded by tall, species-rich hedgerows. There is an orchard with over 250 heritage varieties of fruit. A community-supported kitchen-garden which supplies the FarmED café with delicious fruit and vegetables. FarmED provides for its local community. Businesses, volunteers and teams connect over local food produced alongside nature and at the same time drawing down carbon into its farmland.

At FarmED there are a wide range of educational courses, events and farm walks throughout the year. Farmers and growers visit along with their advisors. So too do university and college students as well as industry leaders, politicians and business executives from the food manufacturers and retailers.

We aim to provide information, inspiration and connections for people wanting to make a difference. To forge a path for people involved in the food supply chain to see what can be done for a future where nature, people and food production are in harmony. It starts with the soil ...

Food, Farming & Countryside Commission

The FF&CC were set up to explore practical and radical solutions to the climate, nature, health and economic crises of our time. Through evidence, research, telling stories of change and much more, we seek to involve and communicate with citizens, and advocate for new ideas and new solutions.

The FF&CC started in November 2017 as an independent inquiry hosted by the RSA (Royal Society for the encouragement of Arts, Manufactures and Commerce) and funded by the Esmée Fairbairn Foundation. We commissioned research and sought practical solutions to tackle climate, health and nature crises. The inquiry formed the basis of a landmark report, Our Future in the Land. We became an independent charity in April 2020.

Our partnerships with leaders across diverse sectors, disciplines and places, help us to liberate new possibilities, shift public narratives, and create the conditions for governments and businesses to propose and implement progressive policies and actions. Our distinctive approach is systems-thinking, collaborative, inclusive and evidence-led, reaching out to seldom heard voices and engaging non-traditional sectors in the work.

We aim to help shape a transformation of our food system so that:

- Healthy food is everybody's business, levelling the playing field for a fair food system.
- Farming is a force for change, with a transition plan for agroecology by 2030, and the resources to back that plan.
- The countryside works for everyone with flourishing rural economies and thriving communities where people can afford to live and work.
- Money flows to where it is needed to align investment with the priorities for a just transition.

In December 2023 we launched Multifunctional Land Use Framework, reporting on our pilot programmes in Devon and Cambridgeshire showing that a Multifunctional Land Use Framework (MLUF) offers the greatest potential to maximise delivery of social, economic and environmental objectives and align resources to provide public value and create a sustainable future.

We Feed The UK is a nation-wide storytelling project led by The Gaia Foundation. It has been co-created as a collaboration with over 20 arts and environmental partners, 10 world-class photographers and 10 award-winning spoken word artists to support the regenerative food and farming transition to reach new audiences. We Feed The UK will run from February 2024 to May 2025 in galleries nationwide, in community venues and in unexpected corners of the UK & Ireland.

We Feed The UK celebrates family farmers and food growers as custodians of biocultural diversity and trailblazers of the grassroots solutions needed to address the environmental, climate and social crises we face. Throughout 2023 environmental organisation The Gaia Foundation forged a network of arts partnerships nationally to co-commission award-winning documentary photographers to tell these stories. Through a collaboration with Hot Poets they added to the storytelling recipe 10 regionally based spoken word artists who are championing bringing fresh voices and audiences to the critical issues of our time. Together through lens and word, they have captured stories of regenerative farming from ten areas of the UK, weaving an inspiring and hopeful narrative for the future of food and farming.

The Gaia Foundation have been flying the flag for regenerative, holistic approaches to protecting and reviving biocultural diversity for almost 40 years. Established in the 1980s in response to indigenous peoples' displacement from forests in the Amazon, Gaia have become a well-respected voice on issues ranging from land rights to indigenous knowledge to seed sovereignty. They run the Seed Sovereignty Programme for UK & Ireland supporting the revival of seed saving and small-scale seed production, and restoring agri-diversity to these isles.

(l) Nikki Yoxall, Grampian Grazers. Photo: Sophie Gerrard. (c) Paulette Henry, Black Rootz. Photo: Arpita Shah. (r) James Robinson, Strickley Farm. Photo: Johannes Pretorius.

www.wefeedtheuk.org

Groundswell, entering its eight year in 2024, provides a forum for farmers and anyone interested in food production or the environment to learn about the theory and practical applications of conservation agriculture or regenerative systems, including no-till, cover crops and re-introducing livestock into the arable rotation, with a view to improving soil health.

It is a practical 2-day show aimed at anyone who wants to understand the farmer's core asset, the soil, and make better informed decisions. With a wealth of talks, forums and discussions from leading international soil health experts, experienced arable and livestock farmers, agricultural policy experts, direct-drill demonstrations and AgTech innovators.

With wide appeal across the food and farming spectrum, Groundswell is relevant for conventional, organic, livestock, arable, landowners or tenant farmers. Groundswell was founded by the Cherry family on their mixed farm in Hertfordshire. John and Paul Cherry have farmed for over thirty years, converting to a no-till system in 2010.

> We started Groundswell out of a sense of frustration that no-one was putting on a Summer Show to which we might want to go. A visit to the fabulous No-Till on the Plains Conference in Salina, Kansas showed us what could be done.

And so, Groundswell began 7 years ago and is growing, and it is going from strength to strength. In 2023 we welcomed over 6500 delegates. This growth reflects the snowballing of interest in regenerative agriculture, not only from farmers but also from policy makers attracted by the 'public

benefits' of such systems such as carbon sequestration and flood/drought prevention. And then there are the Foodies who come to find new and better ways to join the conversation about how and what we might grow in the future; they don't need to be told that food tastes better when it grows in healthy soil.

275

The Nature Friendly Farming Network (NFFN) is a UK farmer-led network working to drive recognition that nature-friendly farming is the most sustainable way of producing food. They are a membership organisation led by farmers, working alongside other organisations and public supporters. They unite farmers across the UK who champion how food and farming can positively influence change. Together, they are a strong voice for sustainable food and farming in the UK.

With 71% of UK land taken up by agriculture, farmers are the most important stewards of our environment. Without them, it will be impossible to meet the UK's nature and climate goals. Right now, in the UK, we are at a critical time both for the future of agriculture and the health of the environment.

The NFFN's work is to support farmers at every stage of their journey towards nature-friendly farming. Through showcasing the experiences of farmers who are leading the way, they share knowledge that empowers those in transition to produce plentiful food. They are committed to shaping food and farming policies that ensure fairer returns for farmers, improved access to sustainable food and greater stewardship of the environment. They share farm-level experiences of whole-farm approaches that restore rural environments, regenerate ecosystems, protect biodiversity, act on climate change and offer greater resilience to natural or economic shocks.

NFFN's mission is to take nature-friendly farming to a new level, making it mainstream.

Pasture for Life is a membership organisation dedicated to promoting the restorative power of grazing animals on pasture. Pasture-fed farming not only positively impacts farmer livelihoods but contributes to biodiversity conservation, human health and animal welfare. By integrating grazing animals into their farming practices, farmers can increase farm biodiversity and profits.

Through nationwide programmes of farmer-led mentoring, farmers receive guidance and support from experienced farmers who have successfully implemented pasture-fed farming methods. These mentors share their knowledge and expertise, helping farmers to navigate the challenges and maximise the benefits of integrating grazing animals into their own farms.

In addition to mentoring programmes, Pasture for Life organises a variety of events that serve as places for knowledge exchange, collaboration and innovation. Members also gain access to an online forum with advice and guidance. There is also the opportunity to actively take part in the governance of the organisation.

Farmers of all types, alongside many others including butchers, chefs, academics, and members of the public who are interested in the potential of pasture-fed farming can join the organisation. There are no prerequisites for farmers to join with regards to the type of farming methods they may currently practise, nor is there any pressure to change. Though, many farmer members find themselves naturally adopting pasture-fed methods due to the evident benefits demonstrated by other farmers, and go on to embark on becoming Pasture for Life certified, in order to fully realise these benefits. Certification guarantees adherence to a high level of farming standards, and as such, may lead to a price premium for certified business products.

By fostering a diverse community that shares an interest in pasture-fed farming, Pasture for Life aims to create a movement for the wider uptake of pasture-fed methods and its benefits, for a sustainable and regenerative agricultural system. The organisation is at the forefront of promoting restorative ways of farming and encouraging individuals to be part of efforts to enhance food, farming and the countryside for the betterment of us all.

Organic Farmers & Growers was founded in the early 1970s as a marketing cooperative for organic farmers. In 1992 OF&G became the first organic body to be approved by the government and today OF&G certifies more UK organic land than all the other UK certifiers combined. OF&G provides support and guidance to businesses in the schemes that we certify to help them remain compliant all year, every year. We also offer support to those considering the switch to organic.

All organic food and drink sold in the UK must meet the UK Organic Regulation. Organic food and farming are unique in that it is the only defined food production approach that is underpinned by clearly defined legal standards. Organic farming at its heart seeks to work with and enhance natural ecosystems and provides one simple, quantifiable route to address the multiple environmental challenges we currently face.

Over the years OF&G certification has grown to support wider sustainable land use. OF&G now offers inspection and certification for the *Pasture For Life* mark, *Compost Certification Scheme*, and the *Biofertiliser Certification Scheme*. OF&G also offers validation and verification for the *Woodland Carbon Code* and the *Peatland Code* schemes.

Recently we produced '*Growing organic – a multifunctional component of English land use policy*' which identifies how a three-fold increase in organic land use area would reduce total agriculture-related GHG emissions equal to the carbon sequestered by a third of million acres of broadleaved woodland.

Aside from the environmental impacts of not applying synthetic fertilisers and pesticides on pollution, air quality and greenhouse emissions from their manufacture and application, by removing artificial inputs, the white paper also highlights biodiversity improvements. In organic systems arable plant species were found to be up to 95% higher, field margin plant species up to 21% higher, farmland bird species increased by 35%, pollinators are up by 23% and earthworm species increased by 78%.

"Don't be scared of certification – it's part of the learning curve". – John Pawsey, organic farmer, Suffolk

The full paper is available to download from www.ofgorganic.org

Contact us at info@ofgorganic.org or call us on 01939291800

The Soil Association is a charity that has been working for over 75 years to transform how we eat, farm and care for our natural world. The only way to solve the issues facing our world is to understand that they are all connected and that food, farming and forestry are vital to the solution.

Over time, the Soil Association has built a fleet of initiatives from the ground up like Soil Association Certification Limited, Food for Life, Exchange, Innovative Farmers, Get Togethers, Sustainable Food Places and the Land Trust to help people build natural solutions together. Through these entities we campaign for change, support farming innovation, serve healthy food, develop world-leading standards, support and grow the organic market, and protect forests. We use our learnings to deliver evidence to the government to support positive changes to food, farming and land-use laws.

We champion progress away from intensive production and consumption systems towards systems that promote health, ecology, fairness, and care – systems like organic, which move us closer to a nature-friendly future and support sustainable living and working.

And we do this work globally. We were one of the founders of the global organic movement and developed some of the world's first organic standards, allowing us to protect producers, consumers, and the soil by endorsing nature-friendly farming methods. This work continues as we push standards for food, farming, health and beauty, textiles and forestry, and work across Europe to influence legislation.

In collaboration with our colleagues across the food, farming and forestry sectors, our expertise allows us to continue transforming how we eat, farm, and care for our natural world. Working together, we will build natural solutions that will help create a nature-friendly future.

Sustain: The alliance for better food and farming is a powerful alliance of organisations and communities working together for a better system of food, farming and fishing, and cultivating the movement for change. Together, we advocate food and agriculture policies and practices that enhance the health and welfare of people and animals, improve the working and living environment, enrich society and culture, and promote equity.

Working in collaboration, we:

- Develop networks of people and organisations to devise and implement projects and campaigns, and to provide a platform for recognition and replication of pioneering work.
- Run highly effective and creative campaigns, advocacy, networks and demonstration projects, aiming to catalyse permanent changes in policy and practice, and to help equip more people and communities with skills as change-makers.
- Advise and negotiate with governments, local authorities, regulatory agencies, funding bodies and other decision-makers to ensure that legislation and policies on food, fishing and agriculture are publicly accountable and socially and environmentally responsible.

Working with our members and supporters, we campaign for a healthy and sustainable food system that is publicly accountable and socially and environmentally responsible. Our campaigns and projects fall into the following seven areas of work: (1) National food strategy; (2) Sustainable farming and fishing; (3) Climate and nature emergency; (4) Brexit and trade; (5) Good food economy; (6) Good food for all; and (7) Local action.

Membership is open to national organisations which do not distribute profits to private shareholders and which therefore operate in the public interest. The organisations must be wholly or partly interested in food or farming issues and support the general aims and work of the alliance.

In the 7 years since its inception, the Sustainable Soils Alliance (SSA) has established itself as the UK's leading soils focused organisation, driving debate around some of the critical agri-environmental issues of our time: regenerative farming, sustainable food production, flood-risk management, pollution-free rivers and soil's role in climate change mitigation.

Our objective is to design and promote government, corporate and farming policies that will achieve the goal of *sustainably managed soils within the space of a generation*. Our work involves answering questions that include:

- How to ensure our soils are able to mitigate the effects of climate change and continue to produce food in the face of extreme weather?
- What regulations are needed to ensure our soils are well managed, do not pollute our streams and rivers, and store as much organic carbon as possible?
- How to ensure farmers are adequately paid to implement practices that protect and improve their soils?

The challenge of answering these questions is shared between policymakers, businesses, academics, NGOs and farmers. To reflect this, we operate as much as a thinktank as a campaigning organisation – convening and aligning these organisations in order to share and amplify their insights and expertise.

As a result of our work, and that of our collaborators, soil health now features in *farming incentive schemes*, an ambitious nationwide monitoring scheme is underway, and soil carbon is starting to appear in government and corporate strategies for climate change mitigation.

Public interest in soil is also growing thanks in part to our work with UKCEH on the UK's first Soil Awareness Week (https://uksoils.org).

Advising farmer profiles

Detailed profiles from the farmers who advised Adrienne, Anna and Ben.

Marina O'Connell
Apricot Centre / Huxhams Cross Farm
Biodynamic mixed farm
Advised Adrienne (see Chapter 5)

John Pawsey
Shimpling Park Farm
Organic mixed farm
Advised Anna (see Chapter 4)

Tom Pearson
Pearson Gape Farming Partnership
Regenerative arable
Adrienne's host farm/landlord (see Chapter 5)

Nic and Paul Renison
Cannerheugh Farm
Regenerative livestock
Advised Ben (see Chapter 6)

FARM PROFILE

YOUR NAME(S)
Marina O'Connell

BUSINESS/FARM NAME
The Apricot Centre /
Huxhams Cross Farm

REGION
Dartington

BUSINESS TYPE
(e.g. mixed farm, smallholding, forestry)
Biodynamic mixed farm

SIZE *(land area owned/managed)*
13 ha

PRODUCTS *(principle business outputs)*
Vegetables, fruit, chicken and eggs,
small-scale grain, cut flowers, cattle
and pigs
Well-being work and training

RAINFALL *(average annual)*
900–1000 mm

ALTITUDE
30-80 m

SOIL
Clay

**YEAR ESTABLISHED AND/OR
TOOK CONTROL/BOUGHT**
2015

ANNUAL AVERAGE TURNOVER
£1 million

EMPLOYEES *(include yourselves)*
20

LEGAL FORM OF OWNERSHIP
(company, sole trader, trust etc.)
Community interest company
(CIC)

TENURE
(of land ownership)
Tenancy with the Biodynamic
Land Trust

YOUR APPROACH
(self-defined, as you see it)
Permaculture

**KEY AGROECOLOGICAL
PRACTICES**
Biodynamics, agroforestry

OVERVIEW

The Apricot Centre moved to Huxhams Cross farm in 2015 from a much smaller site
in Essex where it had been for the previous 20 years. This was a Biodynamic Land
Trust (BDLT) farm, and the founding directors Marina and Mark O'Connell were
interested in taking on this new tenancy with Bob Mehew and Rachel Phillips to
scale up their offer of food production coupled with well-being work and training.

The farm has evolved very much as a team enterprise, with Bob Mehew heading up the finance and business management of the company, Rachel Phillips has created the School for Regenerative Land-based Studies over the last few years, and Mark O'Connell heads up the well-being aspect that is delivered in and round the farm itself, with a team of practitioners who are prepared to work with children on the farm in all weathers. We have a brilliant team of young growers and farmers, who produce great food that is sold and delivered within a 20 mile (32 km) radius of the farm as direct sales.

2015

2019

2023

HISTORY OF LAND/YOUR RELATIONSHIP WITH IT

The 13 ha that is now called Huxhams Cross Farm had been farmed industrially for the previous 40 years by the dairy farm at Dartington Hall Trust. As an outlying piece of land with poor soil it was sold to the BDLT in 2015. The BDLT carried out a share offer and there are 150 investors in the small farm. The Apricot Centre took on the first tenancy. It was a bare field site, with three arable fields, two wetland meadows that had been abandoned and a field that was described as a 'miserable bit of land' by the previous farmer.

We carried out a permaculture design process, involving the BDLT and other interested parties, and created a vision for the layout of the new farm. This was implemented between 2015 and 2017 and went on to became a fully functioning biodynamic farm producing vegetables, eggs, fruit, small-scale grain, and more recently pigs and cattle.

In 2021 we took on the glebe fields next door on a short-term tenancy from the Diocese of Exeter, this has expanded our small farm to 23 ha.

We have embedded a well-being service for adopted and looked-after children into the farm, with areas dedicated for group and individual work, both outside and in small buildings. We also host school visits.

In 2022 our School for Regenerative Land-based Studies was established, and the Apricot Centre now delivers Level 2, 3 and 4 courses in regenerative land-based systems, which are formal qualifications and funded by local government.

MOTIVATIONS

I am a horticulturist, with a degree in what I now call industrial horticulture from the 1980s. After completing my first degree I immersed myself in the world of what we now call regenerative farming systems, learning these methods over many years of practice. This was then backed up by a masters' degree in Environment and Society supervised by Jules Pretty in the early 2000s. I have been passionate about producing food using these methods for many years and can see how they can contribute towards climate mitigation and biodiversity loss. This culminated in writing a book, *Designing Regenerative Food Systems and Why We Need Them Now*, published in 2022.

We have carried out an impact assessment over the first 5 years of the farm. We designed the farm to meet the challenges of climate change, biodiversity loss and to produce healthy food for local people, while still being economically viable. The impact assessment showed that we have achieved this in just 5 short years.

Mark O'Connell is a child and family psychotherapist and he recognised early on that children with high levels of early year trauma benefit from therapies based in nature, and around food. This is currently being explored more with research into 'Food and Mood'.

As a team our motivation is very much driven by climate change, we are training the next generation of regenerative farmers and growers, which will help in a very small way in the transition to regen ag systems. It is, however, a very small contribution and chronically underfunded.

YOUR APPROACH, WHAT YOU DO AND WHY

We manage the land using the biodynamic principles and practices. We have planted thousands of trees in our agroforestry system and we take the approach of the permaculture ethics of 'Earth care, People care and Fair shares'. I developed this combined approach after many years as a practitioner and finding that these systems work really well together with multiple benefits.

WHAT CHANGES HAVE YOU INTRODUCED?

Too much to include here but I can refer you to the case study I wrote up for my book, Chapter 10 'Designing the world we want'. We completely revolutionised the land use on this piece of land basically! See also our website www.apricotcentre.co.uk.

The following is taken from our 2015–2021 impact assessment report.

- The farm sequesters almost 5 tonnes of carbon dioxide equivalent per hectare per year over and above what the farm uses. A total of 64 tonnes per year.
- The soil organic matter went up by 25%.
- The bird species numbers rose by 50%, the orchid numbers rose by 200%.
- The worm count in the soil doubled.
- In 2020, we harvested a total of 15.4 tonnes of fruit and veg. Almost a 20% increase on 2019. In addition, we produced 2.5 tonnes of eggs, 2.4 tonnes of hay (used as animal feed over winter), 6 tonnes of wheat and 2.3 tonnes of straw. This is produced from a cropping area of 5 ha for the vegetables and fruit, and 2 ha for the wheat.
- The turnover of the farm was approximately £200,000 in 2020, with £140,000 of this being from our own produce.
- We employ 6 FTE people on the farm with 3 apprentices.
- On average we have over 300 customers per week within a 12 mile radius of the farm (150 deliveries, 150 plus on the market stall).
- 63% of our customers said that our engagement has helped them be more resourceful with their food.
- In 2020 we had 1420 visitors on the farm.

**WHAT DO YOU PLAN TO CHANGE/INTRODUCE
AND CONTINUE WITH IN THE COMING SEASONS?**

We are looking to take on another small patch of land to bring up the total acreage to about 100 acres, or 40 ha, so that we can grow grain on a scale that is economically viable. We also need to put in a grain storage barn. We work in partnership with a bakery that mills and bakes with the flour and we would like to consolidate this work and bring it in house for the future.

With this extra acreage we can also have livestock, cattle and pigs, to graze on the rotational green manures.

5 KEY TIPS FROM YOUR EXPERIENCE

1. That regenerative farming and systems are remarkably simple and straightforward and work really quickly once you make the decision to transition.
2. That on the whole the people involved in this work are great people to work with and it is much more interesting and fun than industrial farming.

3. It requires perseverance, as things do go wrong, often due to climate change these days, but you need to carry on with it and it gets easier after the transition period during which your farm and soils often have a little 'wobble' and they heal themselves.

4. Regenerative ag practitioners are pioneers and have to adapt practice to their soils and climate and markets, there is no textbook on how to do this but there are fantastic WhatsApp groups that offer support and encouragement and advice, so join as many as you can.

5. Be bold and make the change! It will absolutely work out, economically as well as environmentally, the water will start working again on your farm, insects and birds will turn up, people will love your produce.

FINANCE YOU'VE RECEIVED/ACCESSED

BDLT bought the farm and paid for the barn and a training centre, worth about £350k in total. The Apricot Centre funded the soft infrastructure of trees tunnels and tractors, etc.

CSS and now SFI grants.

Leader funding paid for 50% of the training centre and some set up mill equipment.

LEAP loan and private investor loan for the PV panels on the buildings.

Philanthropic funding for the impact assessment.

Local government funding to deliver the training courses so that they are free for the participants.

RESOURCES YOU'VE FOUND USEFUL

WhatsApp groups - set up by the LWA, Gaia Foundation, etc. These are for the SW grain group, SW land workers, Pastured poultry, regen ag etc. These are peer-to-peer learning and support at their best!

BDLT meetings – 'Family of Farms' is a fantastic small group of farms that meets every year.

Biodynamic Association website - amazing resources.

ORFC - touch base every year.

EVENTS/COURSES YOU HAVE ATTENDED AND FOUND USEFUL OR NOT

I love the ORFC every year and find that this helps me keep up with changes happening.

ORGANISATIONS/GROUPS YOU BELONG TO

Biodynamic Association
BDLT family of farms
Permaculture Association
Informally the Agroforestry networks

PROFESSIONAL ADVICE

From the BDA
Land agents for tenancy agreements
Farm Carbon Toolkit for soil detail and management
Vets of course
(+ I have been doing this a very long time)

RESEARCH

I often say yes to these but to be honest too many to list!

MEASUREMENT/ASSESSMENT

Key findings from the Farm Carbon Calculator (FCC) for 2020.
Total annual carbon emissions 53.16 tonnes CO_2e
Total annual carbon sequestration 117.70 tonnes CO_2e
Total carbon balance −64.82 tonnes CO_2e
Carbon balance per hectare −4.78 tonnes CO_2e
Carbon balance per tonne of product −2.69 tonnes CO_2e

HOPES AND FEARS

Climate change – adapting to climate change is a huge challenge while
remaining economically viable – it is very stressful.

Governments lack of commitment to regen ag, and financial models that do
not support regen ag.

That people cannot afford to even buy any food, let alone access the healthy
food that they need.

ANYTHING ELSE YOU'D LIKE THE WORLD TO KNOW?

It is remarkably straightforward for farming to transition to regenerative
models that can be a key part of climate change mitigation and adaptation,
while bringing local prosperity and bring real joy to local communities.

FARM PROFILE

YOUR NAME(S)

John and Alice Pawsey

BUSINESS/FARM NAME

Shimpling Park Farm

REGION

Suffolk

BUSINESS TYPE
(e.g. mixed farm, smallholding, forestry)

Mixed farm

SIZE *(land area owned/managed)*

650 ha (+980 ha managed)

PRODUCTS *(principle business outputs)*

Grains, legumes and lamb

RAINFALL *(average annual)*

650 mm

ALTITUDE

85 m

SOIL

Hanslope series chalky
boulder clay

**YEAR ESTABLISHED AND/OR
TOOK CONTROL/BOUGHT**

1904

ANNUAL AVERAGE TURNOVER

£1 million

EMPLOYEES *(include yourselves)*

9

LEGAL FORM OF OWNERSHIP
(company, sole trader, trust etc.)

Limited company

TENURE
(of land ownership)

Owners

YOUR APPROACH
(self-defined, as you see it)

Organic, regenerative
mixed farming

**KEY AGROECOLOGICAL
PRACTICES**

Rotation, no chemicals

OVERVIEW

I came home to farm in 1985 and farmed it with my maternal grandfather until
he died in 1989. I farmed conventionally for 10 years but over that time became
concerned about how the farm was performing financially and more so about
how our farming methods were impacting our soils and nature. We converted
part of the farm to organic production in 1999, with the last part converting in
2006. We re-introduced animals onto the farm in 2014 and currently have 1000
breeding New Zealand Romney ewes.

We have also been asked to farm some of our neighbours' farms organically
and currently farm a further 850 ha for seven other proximate farms under
contract farming arrangements.

We are enthusiastic environmentalists and have been involved in various environmental schemes since 1988.

We also have 3 biomass boilers on the farm that heat most of the farm buildings as well as a solar array on our grain store to offset our grain drying.

Our business is also a carbon sink as we sequester more carbon that the business emits.

HISTORY OF LAND/YOUR RELATIONSHIP WITH IT

Before I was involved my grandfather was seen as a progressive farmer using all modern techniques expected of him by the industry which included intensification techniques. Those included increased use of chemicals (fertilisers and sprays), hedgerow and woodland removal. 500 acres of the farm was turned into an American airbase in 1943 which saw the removal of woodland, hedgerows and houses on the site.

MOTIVATIONS

My personal ambition is to farm in a profitable way while impacting less on soils and nature. Having a positive and healthy team, be they family members, non-family members working directly for our business, others working indirectly with us, is incredibly important to us as well.

YOUR APPROACH, WHAT YOU DO AND WHY – MORE DETAIL ON YOUR APPROACH TO MANAGING THE LAND IN YOUR OWNERSHIP/CARE AND HOW YOU CAME TO THIS APPROACH

My approach is to be as inclusive as possible when making decisions on the farm which means giving as much information to people as possible as well as listening to them in terms of their expectations when working within our business. I want our vision to be a shared vision where everyone feels a sense of ownership.

WHAT CHANGES HAVE YOU INTRODUCED?

Really see all of the above.

We grow quite a few novel crops over and above the normal wheat, malting barley, milling oats, beans and a multitude of legumes to build fertility. More recently we have grown quinoa, chia, lentils, clovers, buckwheat, phacelia, heritage wheat varieties and vetches.

We do a lot of on-farm trials looking at improving our organic offer in terms of crop quality but also better outcomes for our contract farms. These are now done by my son Rufus in conjunction with other members of the farming team.

In 2020 we planted 50 acres of agroforestry, which was designed around wood production (woodchip and timber) as well as shade of our sheep and potentially tourism.

We have a converted barn where we host 25+ school visits a year. We also do lots of farm tours for farmers and farming interest groups and local people. We also host conferences and the occasional wedding or party.

HOW HAS THIS WORKED OUT? WHAT WENT WELL? LESS WELL?

We are doing so much it's sometimes difficult to understand what is doing well and what is not, especially with the peripheral/smaller projects. What I can say is that the core elements of the business are delivering in terms of profitability and looking at all the wildlife surveys (we do one a year and the same survey on a rolling basis every 5 years) nature is recovering. We have doubled our soil organic matter over the last 15 years and so I feel that we are restoring our soils.

WHAT DO YOU PLAN TO CHANGE/INTRODUCE AND CONTINUE WITH IN THE COMING SEASONS?

More trial work and more use of our barn.

Continue to develop the Shimpling Park brand!

5 KEY TIPS FROM YOUR EXPERIENCE

1. Observe
2. Listen
3. Encourage
4. Be positive
5. Make it fun

FINANCE YOU'VE RECEIVED/ACCESSED

Basic Payment Scheme
Countryside Stewardship
Sustainable Farming Incentive
Landscape Recovery pilot
HSBC bank loan to buy other shareholders out of the business
HSBC bank loan to buy farmland

RESOURCES YOU'VE FOUND USEFUL

Desktop researching the internet and social media (esp. YouTube and Twitter/X).

EVENTS/COURSES YOU HAVE ATTENDED AND FOUND USEFUL OR NOT

Organic Combinable Crops event, Groundswell, BASE-UK, Wildfarmed, Organic Research Centre, Agricology, Organic Farmers and Growers, Suffolk Wildlife Trust and Organic Arable are all important to our business.

ORGANISATIONS/GROUPS YOU BELONG TO

See above.

I started the Suffolk Wool Towns Cluster a few years ago in conjunction with the Suffolk Wildlife Trust as our facilitator. We have recently been successful in securing a maximum bid in a landscape recovery pilot looking at connectivity of our ancient SSSI and secondary woodlands in a joint bid with our neighbouring cluster the Stour Valley Cluster.

PROFESSIONAL ADVICE

All the above-mentioned organisations have been very useful. We don't have an agronomist at the moment, but we are re-visiting that at present. My accountant and solicitor have been with us for a very long time and know our business extremely well and so are both important sources of advice.

RESEARCH

We do quite a bit of our own research but see above for relevant organisations.

MEASUREMENT/ASSESSMENT

Constant interrogation of business performance, wildlife surveys, carbon calculators, soil testing and on-farm trials.

HOPES AND FEARS

Succession planning is very much at the forefront of our thinking at the moment.

I always hope for a good harvest!

No fears …

ANYTHING ELSE YOU'D LIKE THE WORLD TO KNOW?

See https://shimplingparkfarm.co.uk.

FARM PROFILE

YOUR NAME(S)

Tom Pearson

BUSINESS/FARM NAME

Pearson Gape Farming Partnership /
Manor Farm

REGION

Cambridgeshire

BUSINESS TYPE
(e.g. mixed farm, smallholding, forestry)

Arable

SIZE *(land area owned/managed)*

470 ha (420 ha cropped)

PRODUCTS *(principle business outputs)*

Combinable crops

RAINFALL *(average annual)*

560 mm

ALTITUDE

55 m

SOIL

411d Hanslope series
(chalky boulder clay)

**YEAR ESTABLISHED AND/OR
TOOK CONTROL/BOUGHT**

2016

ANNUAL AVERAGE PRODUCTION

Approximately 3000 tonnes
of produce per year

EMPLOYEES *(include yourselves)*

2 FT, 1PT (additional 4 PT at harvest)

LEGAL FORM OF OWNERSHIP
(company, sole trader, trust etc.)

Partnership

TENURE
(of land ownership)

87 ha rented, rest owned

YOUR APPROACH
(self-defined, as you see it)

Regenerative agriculture

**KEY AGROECOLOGICAL
PRACTICES**

No-till, 100% cover/catch cropping,
companion cropping, integrated
livestock, significantly reduced
artificial inputs, no insecticides,
beneficial habitats, use of local
organic manures and compost/
biologicals, shortening supply chains,
community connections, education

OVERVIEW

We produce a range of combinable crops: milling wheat, soft wheat, Weetabix wheat, Wildfarmed milling wheat (first season 2023/24), winter barley (for local pig farm), spring malting barley, beans, milling oats, oilseed rape (OSR). Introducing apple tree agroforestry (180 trees in the ground by spring 2024 with plans for 1000+ apple and nut trees). We rent out 1 ha of land to Sweetpea Market Garden (vegetables) and host a flying flock of 500–800 sheep grazing 25–50% of the farm each year between July and March.

HISTORY OF LAND/YOUR RELATIONSHIP WITH IT

Land has been in the family since 1682 but only farmed in hand since my grandfather's time (late 1950s). Prior to this it was split among multiple tenants on life tenancies. Farmed conventionally by my grandfather and father but with attention to detail and quality environmental schemes. I took over after a 1–2-year transition period between 2015 and 2016. Prior to that I trained as a medical doctor graduating in 2001 and working in hospitals, general practice and public health in the UK and abroad.

MOTIVATIONS

Almost 20 years of practising medicine has no doubt shaped my vision for our family farm with health and wellness at the centre. I want to see the staff I employ flourish. I want to see the local community we live in become better connected to their environment and food. I want the farm to offer ways for them to do that. I want the health of our land to improve both above and below the ground.

But with my public health hat on, I need to see this happen beyond the boundaries of our farm. I don't want to shy away from the all too real climate and nature crisis that farmers have the opportunity to make a significant contribution to reversing. A third crisis, highlighted through my medical background, is that of chronic disease, and in particular our broken food system as a cause of this. So ultimately, I want to see real improvements in planet and human health and this needs to go beyond what we do on our farm.

I am an advocate of shared learning: case studies of best practice including the story of how to get there and what not to do. On a local scale I chair our farm cluster group (West Cambridgeshire Hundreds) and aim to help create a group that has confidence to improve their land through group learning, discussion and technical support, with best practice 'on-the-ground' case studies leading to collaborative landscape scale nature projects.

Beyond the neighbouring farm level, I would like to see a thriving local and regional food supply chain offering an alternative consumer option to the globalised system that currently dominates. This should include supplying a significant proportion of public institution catering, particularly hospitals and schools, with healthy nutritious food that has a proven, transparent provenance that is truly planet friendly.

YOUR APPROACH, WHAT YOU DO AND WHY

As a family run (and owned) farm, we are lucky to be able to take a very long-term view. This gives us the freedom to invest resources into actions that might take decades to come to fruition.

We have created a vision and values document that stresses the desire to create positive social and environmental impact. We use these core values to help take business decisions and create long term goals.

WHAT CHANGES HAVE YOU INTRODUCED?

Introduce the **farm mantra of 'Living roots and diversity'**. Great to have an overarching reminder of what we want our soils to have.

Stopped using insecticides after my first two seasons (inspired by a 100% resistance to pyrethroids result on cabbage stem flea beetle (CSFB) [Rothamsted sample from our farm]).

Increasing beneficial insect habitats.

No-till with goal of **100% cover and catch cropping** – phased approach moving farm into this system in blocks of approx 80–100 ha per year (after 3 years of experimenting on 30 ha).

Companion cropping including OSR with berseem and buckwheat, spring oats and beans ('Boats'), and more recently a microclover understory and winter wheat with winter beans.

Low pressure tyres and CFT (controlled traffic farming) to reduce compaction – particularly important in no-tilling clay soils.

Best practice **bunded sprayer washdown zone with biofilter and a sprayer with PWM and single nozzle shutoff** – if you need to use some pesticides then it is good to manage these as responsibly as possible (Countryside Stewardship 30-40% capital grant on the sprayer zone with biofilter).

Introduced sheep (flying flock owned and managed by independent shepherd).

Started growing some **landrace varieties** (Wakelyns YQ) with very low inputs – great to understand what you can grow without artificial inputs as a baseline to reflect on the other cereal cropping on the farm. Hoping to mill this and sell locally.

First year of **growing for Wildfarmed** – allows us to grow wheat with almost no artificial inputs and know where we are selling to. Another tool to give the below and above ground nature the best chance to regenerate while still growing nutritious food.

Hosting a market garden – Sweetpea Market Garden (just finished its second season, see Chapter 5). Such an eye opener to what your land can produce. Fits in well with our philosophy of ever more diversity. Really fulfilling to get a new entrant farmer/grower on the land.

Agroforestry – started with just a single strip of 56 apple and pear trees within the market garden (really fun plating day involving local community). Next two strips planted winter of 2023/24 after establishing the wildflower strips in the field. Market garden gets to use the fruit as long as they look after the trees. Hoping to continue the strips for the rest of the 15 ha field and then look at nut-based agroforestry (probably hazel and walnut) on an 18 ha field. Great way to increase your carbon sequestration, increase your diversity and increase your climate resilience.

Encouraged as much **volunteer/research surveying of biodiversity** as possible – RSPB (volunteer monitoring of farm wildlife (VMFW) pilot), H3 (WP3 – Landscape scale regenerative agriculture change (h3.ac.uk)).

Financial benchmarking – working not just in a regenerative agriculture benchmarking group but also a group of local conventional farmers with similar soil type. This allows me to see where I sit within regenerative farming businesses but also if I am dropping behind on produce or financial output compared to my conventional neighbours. So far, I tend to sit in the middle on yield but with considerably less inputs.

School education visits – started to get into these, working alongside The Country Trust. Encouraging all the team to be involved. Helps to receive approx. £300 per visit through Defra's Countryside Stewardship 'educational access' offer.

HOW HAS THIS WORKED OUT? WHAT WENT WELL? LESS WELL?

- **What went well:**
  - The slow start worked for us. Particularly as well-established staff needed time to adjust and 'see with their own eyes' (or retire!).
  - Bringing in a young team whose priority was to work on a regenerative farm.

- Ignoring people who said it couldn't work on clay soils! (multiple conversations around this in 2015).
- Working with agronomists who believe that regenerative farming is possible!*
- Best winter wheat yield on the farm in the past 10 years has been on a no-till field. This inspires us to know it can be done.
- Cover crop establishment – so satisfying to see a really good diverse cover crop in late autumn.**
- Drilling 'into the green' in spring is probably the most satisfying part of the way we farm: literally no break in the 'keep living roots in the soil as much as possible' principle.
- Lucky to have started our microclover establishment in a wet year – gives us immediate confidence it can be done.
- Finding a shepherd who gets it!***

- **What went less well:**
  - Although aware of it, not taking the time to prioritise the role good drainage has on clay soils in no-till scenarios. Late to complete ditch maintenance and field drain repairs. Not on top of mole draining and targeted subsoiling.
  - In-field trials … doing them, but not following up or necessarily formally presenting the data to anyone.
  - I had grand plans to have really quality base-line data and am far from the complete set I wanted. The 'How to measure soil health' question continues to be debated and there was no consensus/guidelines at all in 2015.

*   Working with agronomists who do not have a long-term game plan for the farm. You need to be reactionary to the seasons, but you also need to continue to move in a direction that promotes continuous regeneration and ongoing reduction in artificial inputs.
**   Cover crop establishment! Slow to understand the significance of herbicide half-lives, straw (residue) management and potential impact of the later fungicide applications.
***   Initially using a shepherd who didn't want to learn a different way of grazing. This led to several years with very bad spring crops due to over grazing in the winter (half the usual yield).

**WHAT DO YOU PLAN TO CHANGE/INTRODUCE
AND CONTINUE WITH IN THE COMING SEASONS?**

Start composting and using 'biologicals' on the farm.

Expand our agroforestry system and make sure we measure the impact and share the story with our local famers.

I would like to consider converting a portion of our farm into organic. I feel it would be good to compare these methods to understand the pros and cons of each. I also want the team to be confident to grow organically if future pesticide licensing or evidence pushes us that way.

More local supply chain products – adding combinable crop produce to the variety of produce available to the market garden customers. This includes getting a mill and working with local bakers.

Develop a solely regenerative flock of sheep and market this (in collaboration with shepherd).

More education sessions for schools.

Work to help develop local supply chain infrastructure and help to develop 'farm to institution' supply.

Learn of, and adopt, systems of equitable food access.

Consider ways of locally building food confidence (preparing, cooking, storing and eating more healthy food) and measuring the health outcome of this.

Get better data on our produce's 'planet footprint'

Study the nutritional value of our produce and compare.

5 KEY TIPS FROM YOUR EXPERIENCE

1. Talk to people who have done it, then get on and try (even just a small area!)
2. Be wary of reading too much into yield/economic comparisons of conventional vs regenerative – you're investing in a 5–10-year plan of transition to deliver more resilient soils (your biggest asset) and rationalising your fossil fuel reliance – the value of this has a place on the spreadsheet!
3. Work tirelessly to get everyone you work with on board and don't be afraid to say goodbye to the naysayers.
4. Trust good science but don't use a lack of research as an excuse to not try yourself.
5. Record your story, share your experience, encourage and support others not to reinvent the wheel or fall into the same traps you have.

FINANCE YOU'VE RECEIVED/ACCESSED

Defra BPS (but this is being phased out).

Defra ELS/HLS scheme and from January 2023 a new Countryside Stewardship (CS) scheme, also including some SFI payments from April 2024.

Defra grants from CS-mid tier (concrete and sustainable drainage system for yard: 30–40% grants, 2 km gapping up hedgerows).

Defra Farming Equipment and Technology grants (approx. 40%) (rainwater harvesting, direct drill).

Bank overdraft and loans.

Occasional 0% finance on new machinery.

It is important to note that we are lucky enough not to be chasing cash flow on a monthly basis, and although we can't afford major upsets for multiple years, we do have some wiggle room to experiment and push some agronomy and system boundaries.

RESOURCES YOU'VE FOUND USEFUL

Groundswell – great buzz, melting pot of people, experience and ideas.

BASE-UK – great to know there are other farmers out there trying the same system and pushing the boundaries.

Farmarama podcast – beautifully collated, people focused.

Local WhatsApp group for no-till farming (which I jointly set up).

Books: Gabe Brown – *Dirt to Soil*, Nicole Masters – *For the Love of Soil*, David R. Montgomery & Anne Biklé – *What Your Food Ate*, Tim Spector – *Food for Life*, Sarah Langford – *Rooted*, Henry Dimbleby – *Ravenous*, Chris Van Tulleken – *Ultra Processed People*.

Current Nuffield Scholarship is giving me the opportunity to visit and talk to practitioners/experts in the food system, opening my eyes to alternatives to the status quo.

EVENTS/COURSES YOU HAVE ATTENDED AND FOUND USEFUL OR NOT

Groundswell (as mentioned above) – so helpful to see so many people taking a similar agricultural journey to me.

Joel Williams' Foliar Nitrogen Course.

Farm visits with discussions.

I'm increasingly becoming more detached from the usual events (Cereals, LAMMA, etc.).

ORGANISATIONS/GROUPS YOU BELONG TO

Red Tractor assurance, NFU, West Cambridgeshire Hundreds (WC100s) Group (farm cluster, chair), Green Farm Collective, BOFIN, NIAB.

PROFESSIONAL ADVICE

Agronomist – Edaphos.

Attend online talks and short courses specifically around agroecological and regenerative farming, especially ones that help us use techniques to reduce artificial inputs; for example, Joel Williams' Foliar Nitrogen Course.

RESEARCH

The H3 Programme (Healthy Soils, Healthy Food, Healthy People [h3. ac.uk]): we are part of 'work package 3', a 4-year research project working at the landscape scale, evaluating the impact of regenerative agriculture in terms of soil health, wider environmental outcomes and food quality. I sit on a fortnightly meeting, steering the operational direction of the research and advising from a farmer's perspective.

We tend to **host one to two trials** on the farm each year **- BOFIN** – slug resistant wheat trials, **Rothamstead** – CSFB, parasitic wasp trials, **NIAB** – wild oat control, **ADAS** – OSR seed rate trials, **KWS** – no-till seed trials.

I attended the **Silvoarable agroforestry ELM test** workshop (2021) and advice pilot (2024).

I am part of the FFCC leadership group piloting a land use framework for Cambridgeshire.

We have just started a 'compost club' within the WC100s cluster membership to learn from each other about different composting techniques and how to integrate them into our farming practices.

I am currently undertaking a **Nuffield Faming Scholarship** that asks: *'Can farmers and growers make a positive impact on local community health?'* This gives me the opportunity to look at systems that deliver equitable access to planet friendly nutritious food while building food confidence and measuring health benefits to this.

MEASUREMENT/ASSESSMENT

We use **Vidacycle's soil mentor regen platform** – this helps record in-field tests that give indications of improving soil health.

We are starting to use the **Healthy Hedgerows** app – this allows us to record the state of our hedges, the first step of improving these vital nature assets on our farm.

We are part of the **RSPB Volunteer Monitoring of Farm Wildlife project** – volunteers come to farm and undertake several bird, pollinator and butterfly surveys throughout the year.

We do regular soil lab testing including some **carbon stock testing** – this allows us to understand the baseline 'stock' of carbon we have stored in our soils at 30 and 60 cm depths. Also **soil audits** to understand cation exchange and nutrient holding and releasing capacity of our soils.

We do some tissue testing and are hoping to use more **Brix testing** as well as pH, chlorophyll and sap testing.

We are creating a **cluster-wide map (using the Land App)** of current habitats and overlaying 'opportunity mapping' of potential additional habitats. This allows us to make intelligent nature decisions that lead to maximum interconnectivity of nature between farms.

HOPES AND FEARS

What encourages me most:

We are getting annual improvements in our regenerative techniques and I have a team on board that is invested in making it work.

Our soil has shown it can recover and regenerate, given the right conditions. We are fortunate to have a clay soil (but not too much clay) that is self-structuring and can hold on well to organic matter and nutrients.

We have a domestic agricultural policy that, although not perfect, does understand the need to move to more nature friendly regenerative farming techniques and has a system coming into place that financially helps farmers move that way.

What concerns me the most:

Will the political will remain to continue funding a forward-thinking planet friendly domestic agricultural policy?

Climate changes that make it more and more risky to farm … very wet winters and dry springs result in failed winter crops (waterlogged) and failed spring crops (drought). Our regenerative practices should result in more resilient soils but how much can even they take?!

Transitioning to regenerative agriculture takes time, exposes the farm to potential short-term risk (for long-term gains) and needs considerable change in mindset and additional skills. Do farmers and growers have the capacity to stay on that journey?

The time to make major positive changes to climate and nature friendly farming and food systems was two decades ago. Are we too late?

ANYTHING ELSE YOU'D LIKE THE WORLD TO KNOW?

I truly think that all people need to become more connected to their food. Understanding where food comes from and how it is grown gives people an appreciation of the value of that food and equips them to make informed choices that consider the total planetary cost of that food, as well as its nutritional value and health implications. This is a multifaceted problem that requires a complete rethink to our food, education, infrastructure, health and social care systems. We need a system that supports equitable access to healthy nutritious food, and in doing so embraces food nutrition knowledge and food confidence (choosing, preparing, cooking and storing food and managing food waste). Farmers and growers can play a unique role in this change.

REFERENCES

Wildfarmed (https://www.wildfarmed.co.uk/farming)
The Country Trust (https://www.countrytrust.org.uk/)
BOFIN (https://bofin.org.uk/)
Vidacycle's Soil Regen Platform (https://soils.vidacycle.com/soil-tests/the-regen-platform/)

FARM PROFILE

YOUR NAME(S)
Paul and Nic Renison

BUSINESS/FARM NAME
Cannerheugh Farm

REGION
Cumbria

BUSINESS TYPE
(e.g. mixed farm, smallholding, forestry)
Livestock

SIZE *(land area owned/managed)*
141 ha

PRODUCTS *(principle business outputs)*
Meat

RAINFALL *(average annual)*
1206 mm

ALTITUDE
200–335 m

SOIL
Sandy loam

**YEAR ESTABLISHED AND/OR
TOOK CONTROL/BOUGHT**
2012

YOUR APPROACH
(self-defined, as you see it)
Regenerative

**KEY AGROECOLOGICAL
PRACTICES**
Mob grazing, long rest period,
no fertiliser/chemicals

OVERVIEW

Reno (Paul) and I have farmed here since 2012, I'm from a dairy farm on
the Welsh border and Reno is from the Wirral (non-farming background).
We arrived here with a very conventional mindset, Reno had spent 10 years
managing a fell farm in the Lake District and I had milked a lot of cows and
worked for agri-pharma/management companies. We quickly realised that
conventional farming relies on huge investments in kit and inputs, we then met
some farmers who were doing things differently at scale and pretty much since
then have been doing things differently. Slowly at first, building up confidence,
realising that mistakes are ok.

MOTIVATIONS

We are both motivated by the huge benefits that 'farming with nature' has for
both the environment but also for the profitability and long-term prospects
for both our business and our planet. I have worked in knowledge exchange

previously and having farmers visit us is now quite a big part of life, we also set up Carbon Calling (google it!).

YOUR APPROACH, WHAT YOU DO AND WHY

We produce 100% pasture-fed beef, no fert/sprays/chems are used, cattle stay out for 9 months of the year, low input system, mob grazing, long rest periods.

Cows are followed by 'pastured laying hens', these hens are truly free range, alongside the forage they are fed soya free grains.

WHAT CHANGES HAVE YOU INTRODUCED?

We didn't get much baseline info when we started – big mistake!

We have just had a bird survey done: we had 52 breeding bird species on the farm and all five uk owl species breeding, which on a UK upland farm is good (so they say!).

Longer rest periods mean more insects, more birds – more life.

HOW HAS THIS WORKED OUT? WHAT WENT WELL? LESS WELL?

We are now making money at the same time as holding water, increasing bird/insect life/soil health – I don't think we would be in the conventional system.

WHAT DO YOU PLAN TO CHANGE/INTRODUCE AND CONTINUE WITH IN THE COMING SEASONS?

More trees, more data collecting.

5 KEY TIPS FROM YOUR EXPERIENCE

1. Wean yourself off fertiliser slowly over 3 years.
2. Read/listen to books.
3. Understand how grass likes to grow.
4. Dig holes.
5. Work out your cost of production.

FINANCE YOU'VE RECEIVED/ACCESSED

Higher Level Stewardship (HLS), Countryside Stewardship (CS), Farming in Protected Landscapes (FiPL).

RESOURCES YOU'VE FOUND USEFUL

Books: Joel Salatin/Richard Perkin/Nicole Masters/Jim Gerish (see Further reading).

Visiting other farmers in UK.

EVENTS/COURSES YOU HAVE ATTENDED AND FOUND USEFUL OR NOT

Carbon Calling, Groundswell

HOPES AND FEARS

I hope as an industry we wake up and smell the coffee, rather than being led by green washing professionals, i.e. Big Ag.

ANYTHING ELSE YOU'D LIKE THE WORLD TO KNOW?

Let food be thy medicine!

Contributors

Mike Berners-Lee is an English researcher and expert on carbon foot-printing. He is a professor and fellow of the Institute for Social Futures at Lancaster University and director and principal consultant of Small World Consulting, based in the Lancaster Environment Centre at the university. His books include *How Bad Are Bananas?*, *The Burning Question* and *There Is No Planet B*.

Matt Chatfield, after spending time away from his family's Devonshire farm working in London at a publisher, decided to return home and set up The Cornwall Project. The aim was to drive interest in Cornish produce and help set up connections between suppliers and restaurants. Over the years he acted as a middleman between suppliers, including Philip Warren Butchers and top London restaurants such as Kiln, Brat and Bao. These connections he created between Cornwall and London are still strong today and Matt was instrumental in helping Ian Warren set up his celebrated 'On the Pass' range of chef-inspired aged specialist cuts during lockdown. The past 5 years have seen Matt shift his focus back to the family farm, where he's doing something pretty special with sheep.

Henry Dimbleby, MBE, was appointed lead non-executive board member of the Department for Environment, Food and Rural Affairs in March 2018. He was co-founder of the Leon restaurant chain. He is also co-founder of The Sustainable Restaurant Association, Chefs in Schools and Bramble Partners. He co-authored two independent reviews for government: *The School Food Plan* (2013), which set out actions to transform what children eat in schools and how they learn about food; and the *National Food Strategy* (2021).

Molly Foster (researcher, *Six Inches of Soil*) is an interdisciplinary researcher whose interests relate to food and farming sustainability. She studied human sciences at Oxford University and is currently pursuing an anthropology PhD at the Natural Resources Institute, University of Greenwich. She has previously carried out research into corporate venturing into alternative proteins, with the University of Oxford's Smith School of Enterprise and the Environment, and taught undergraduate human ecology.

Adrienne Gordon is a new-entrant farmer who recently set up Sweetpea Market Garden on 4 acres of rented land in Caxton, Cambridgeshire. A love of cycle touring took her to New Zealand where she volunteered at organic farms and community gardens before embarking on a traineeship at a 1/4 acre market garden. She moved to Sussex in early 2020 to work at an organic vegbox scheme, and subsequently managed a 140 share CSA with the team at Pea Pod Veg. She's now returned to her roots in Cambridgeshire to be closer to family and to embark on this next big adventure!

Vicki Hird was, until recently, Head of the Sustainable Farming Campaign for Sustain: The Alliance for Better Food and Farming. She is now Strategic Lead on Agriculture at the Royal Society of Wildlife Trusts and she also runs an independent consultancy. An experienced and award-winning environmental campaigner, researcher, writer and strategist working mainly in the food, farming and environmental policy arenas, Vicki has worked on government policy for many years and is the author of *Perfectly Safe to Eat?: The Facts on Food* and *Rebugging the Planet*. Vicki's passion is insects. The first pets she gave her children were a family of stick insects, and she received a giraffe-necked weevil tattoo for her 50th birthday. Vicki has a masters in pest management and is a fellow of the Royal Entomological Society (FRES).

Andrew Jackson began farming, in 1975, with his father and they practised conventional farming with ploughs and chemicals. In 1997, he transitioned 160 acres (65 ha) to organic production, accompanying it with a small farm shop. In 2010, they left organic production due to profitability challenges while supplying supermarkets, transforming the farm shop into the Pink Pig Farm, a farm attraction with a restaurant. Andrew's journey toward regenerative farming started in 2015, involving joining BASE-UK and adopting no-till practices in 2020 while reducing inputs every year. Recently, they established a cluster group to support local regenerative farmers in North Lincolnshire.

Anna Jackson's life journey is a compelling blend of urban and rural experiences. She started on a farm but spent 7 years as a photographer in Oxford and London. During the last 3 years, she launched her sports photography business while running 'A Zero Waste Life', a non-profit dedicated to waste reduction. The COVID-19 pandemic prompted her return to the farm, which turned out to be a fulfilling decision. Inspired by her father and influenced by *Dirt to Soil* by Gabe Brown, she embraced regenerative farming. Their farm focuses primarily on arable cultivation, utilising sheep for grazing and disease control. They haven't ploughed since 2015 and adopted a no-till approach in 2020. Crops include peas, 10-way wheat blend, 3-way oilseed rape blend with clover, grass with clover, oats, and beans. They also rent land for pigs to create a more sustainable rotation.

Hannah Jones is a soil and carbon advisor with the Farm Carbon Toolkit. She works with a range of farmers and farms businesses to support profitable transition to reducing greenhouse gas emissions. Much of her work focuses around optimising farm resilience through improved soil health and management.

Priya Kalia (communications advisor, *Six Inches of Soil*) is a biologist with over 12 years' experience in regenerative medicine research, where she studied how to repair the body when it's ill, injured or ageing. During her time in research, she learned about how food and its micronutrients can positively affect our bodies. Priya studied Molecular Biology and Genetics at the University of Toronto, Canada, has a PhD from University College London, and carried out postdoctoral research at the University of Cambridge and Kings College London. She runs a strategic communications and content advisory firm, Scitribe, where she works with organisations across science, the environment and technology.

Satish Kumar is an Indian British activist and speaker. He has been a Jain monk, nuclear disarmament advocate and pacifist. Kumar is Founder of the Schumacher College international centre for ecological studies, and is Editor Emeritus of *Resurgence & Ecologist* magazine.

Tim Lang, Professor Emeritus of Food Policy, Centre for Food Policy, City University of London, has spent 45 years in public and academic research and debate on food systems and change. After a PhD in social psychology at Leeds University 1970–73, he became a hill farmer in the 1970s in the Forest of Bowland, Lancashire which shifted his attention to food policy, where it has been ever since.

Sarah Langford was for nearly a decade a criminal and family barrister both in London and around the country. She went on parental leave to have her two children and wrote her debut book, *In Your Defence: Stories of Life and Law* published 2018. In 2017 she, with her husband, son and baby, moved by accident from London to the Suffolk countryside, taking on the running of her husband's small family farm. Her story is woven around the stories of other farmers she met in her second book, *Rooted: Stories of Life, Land, and a Farming Revolution* published in 2022. Sarah is currently studying a Graduate Diploma in Agriculture at the Royal Agriculture University.

Claire Mackenzie (producer, *Six Inches of Soil*) produced the original film *From the Ground Up*, for Carbon Neutral Cambridge. She has a background in TV production and charity fundraising and event organising. She has set up environmental groups both in London and Cambridgeshire. With a passion for holistic healthcare she has been a trained therapist for 15 years, and a trustee at Cambridge Cancer Help Centre.

Nicole Masters is a globally recognised agroecologist, speaker and author. She has over 22 years' practical farming and food production experience, with 18 years' experience as a regenerative ag coach and educator. Nicole is formally trained in soil science, organisational learning, pattern thinking, and adult education. She has worked closely with diverse production sectors: dairy, sheep and beef, viticulture, compost, nurseries, market gardens, racing studs, lifestyle blocks to large-scale cropping. Working with such diverse clients has fostered a broad knowledge and understanding of the challenges facing different production systems.

Josiah Meldrum co-founded of Hodmedod, a Suffolk company whose aim is to encourage us to grow and eat a wider range of British grown pulses, grains and seeds – creating healthier and more diverse diets and farming systems. Working with farmers Hodmedod has pioneered 'new' crops for the UK, such as lentils and chickpeas and revived long-forgotten staples, like naked barley and carlin peas; central to this has been finding engaged markets to support a committed group of farmers. This enables change by encouraging the creation of more complex rotations, and through more direct routes to market for what might otherwise be anonymous commodity crops, bringing additional economic and agroecological value to farms. Prior to Hodmedod, Josiah worked for NGOs, farmers' cooperatives and retailers whose aim was to create a more equitable and sustainable food system.

Lucy Michaels (researcher & impact producer, *Six Inches of Soil*) is a scholar-activist with 25 years' experience researching the global food system. She was agribusiness researcher for Corporate Watch UK for 5 years and has run major research projects for Greenpeace International, Compassion in World Farming, World Animal Protection, Global Justice Now and Changing

Markets Foundation, among others. Lucy is an anthropologist by training. During 2020–2022 she was based at Warwick University exploring human–soil relations. She is currently a UK Food Systems Transformation Fellow at the University of Hertfordshire trying to get the UK to eat and grow more pulses. Lucy is the author of *Hungry Corporations*, *A Rough Guide to the Farming Crisis* and *What's Wrong with Supermarkets?* as well as academic articles on climate change and transboundary water management. She has lived and gardened in the UK, US and Middle East.

Emma Mills grew up travelling the world with wildlife and conservation in her blood and homecooked food with fresh ingredients on her plate. She has worked as a partner on the family farm in Wales which harnesses the benefits conservation work can bring to a working farm. Emma co-founded The Kitchen Garden People in order to get nutritious locally grown food onto local plates while changing the landscape of farming and growing into one that works alongside nature and celebrates all the elements of life she holds dear: diversity, real taste and variety.

James Murray-White (researcher, *Six Inches of Soil*) is a multimedia artist and filmmaker, with a background in theatre direction and playwriting. He holds postgraduate degrees in both Human Ecology and Digital Media. He worked for 5 years as an environmental journalist and filmmaker in the Middle East. He was senior editor (culture) at Cambridge TV for two years. James has made films for a variety of platforms, from arthouse to film festivals, online and art galleries including a Channel 4 commission on the life of Bedouin in Israel's Negev desert and a triptych of film-poems with George Szirtes, screened at the Venice Biennale in 2019. James has completed his first documentary feature film *Finding Blake* (2021) which is due for release later this year. James is co-founder of the 'Save the Oaks' campaign, which planted 160,000 broadleaf native trees in the UK during lockdown.

Marina O'Connell is a successful grower, farmer and educator. Marina, the Apricot team and her family have turned the bare land at Huxhams Cross Farm, Totnes, Devon, UK, from being 'a miserable bit of land' as a local farm contractor called it in 2015, into a productive, beautiful, community connected and profitable farm. She has researched the use of biodynamic, organic, permacultural, agroecological, regenerative and agroforestry methods in her work. People from all over Britain and the world visit her farm and go on her courses and a Devon apprenticeship scheme for regenerative food systems.

John Pawsey converted his farm at Shimpling to organic production in 1999 for business reasons, as well as concerns about soil health and decline in farmland wildlife species. John has managed to halt and, in some cases, reverse wildlife decline, improve soil health and keep the farm profitable since that initial conversion. Since then John has increased his organically farmed area from 650 ha to 1,760 ha delivering similar results to neighbouring farmers.

Tom Pearson has for past 9 years been running his family arable farm in Cambridgeshire, transitioning to regenerative practices. He recently started hosting a market garden. He is also a medical doctor, with a masters' in Public Health, and has been practising medicine in hospitals, primary care and public health settings since 2001, both in the UK and abroad. He has first-hand experience of working with patients and families struggling with diet-related chronic disease, and the barriers they face to overcome these or avoid them in the first place. He is fortunate to understand the language of both agriculture and health. He wants to make a significant contribution to bridging the gap between agriculture and health in the UK. He is currently undertaking a Nuffield Scholarship under the title 'What can farmers do to make a positive impact on the health of their local community?'

Colin Ramsay's (director, *Six Inches of Soil*) films have featured in and at *Nature* magazine, BBC click, Cambridge Science Festival, The Festival of Ideas, Institut Français, The Royal Institution of Australia and international film events. In 2017, Colin won the Arts and Humanities Research Council (AHRC) 'Best Research Film of the Year' for his documentary film *Pain in the Machine*. He is the founder of DragonLight Films a production company in Cambridge. He is passionate about nature-based solutions to climate change and sticking it to the man.

Nic Renison farms with her husband Reno in Cumbria. The daughter of dairy farmers, Nic grew up within the conventional, high production ag environment, growing food with little thought of the environment. It wasn't until 2012 when the Renisons started farming their own land they started to think more regeneratively, this wasn't because they wanted to save the world it was because they needed to pay the bills! The last 10 years have been a journey of both practical 're-learning' and a total change in mindset, they now farm 80 suckler cows, with laying hens following the cows around in an 'egg mobile', also a handful of woodland pigs. She is co-founder of regenerative farming conference Carbon Calling.

Lizzie Sagoo is a soil scientist at ADAS with specialist knowledge of soil and nutrient management and diffuse pollution of air and water from agricultural systems. She joined ADAS in 2003 and is Associate Managing Director of the ADAS Sustainable Agricultural Systems Business. Lizzie is a member of the British Society of Soil Science, a STEM ambassador, and a soil champion for the Country Trust's 'Plant your pants' campaign.

Ben Thomas grew up on his family farm in Cornwall. After 3 years at Harper Adam University studying agriculture he then went on to work for an extensive grass based dairy farm and a large mixed farm. Now back in Cornwall Ben manages a herd of Belted Galloways conservation grazing at Goss Moor NNR. This led to Ben and his wife Claudi additionally taking on a small farm at Warleggan on Bodmin moor, using mob grazing to improve soil health, biodiversity, and fatten Belted Galloways and North Devons, to supply premium butchers.

Jeremy Toynbee had his hands in the soil studying archaeology at Trinity College, University of Wales. He then followed a love of books and reading into publishing. As an editor, he has been polishing the words of others and shaping them into books for over 25 years. Following stints in newsstand magazines and with the academic publishers Routledge and SAGE, Jeremy formed his own publishing services company in 2007. He was consulting for 5m Books when the opportunity arose to buy it in 2020.

Ian Wilkinson studied farm and grassland management at Berkshire College of Agriculture and 35 years ago he joined Cotswold Seeds, a family business based in Moreton-in-Marsh, becoming managing director in 1998. Cotswold Seeds has built its reputation on developing forage, herbal leys, green manures and complex seed mixtures. Increasingly, the company acts as a bridge between farmers and the scientific community. Honeydale Farm is a 107 acre (43 ha) demonstration farm in the Cotswolds. It is the home of FarmED.

Dee Woods is a food and farming action-ist who advocates for good food for all and a just, equitable food system, challenging the systemic barriers that impact marginalised communities and food producers. Dee's work sits at the nexus of food and farming, particularly in intersectionality, diversity, equity and anti-oppression, decolonisation, reparations, the right to food and nutrition, participatory policy making, community food systems, food system change, food commons, agroecology, and food sovereignty.

Acknowledgements

The Six Inches of Soil team wishes to thank David White, one of the Cambridgeshire regenerative farmers featured in Colin and Claire's first film, *From the Ground Up*, his farm really opened their eyes to the possibilities. And a massive thank you to all the people that donated to the project and who have supported it so amazingly through its development and since the film's release.

5m Books wishes to thank Agroecology Europe, Cambridge University Press, Hawthorn Press, Rothamsted Research, the Soil Association and Sustain for permission to reproduce the figures, tables and text credited to them in the text. The book would not have been possible without the talents and flexibility of our freelance team, who coped with a moveable feast of a schedule and very tight deadlines: William Newitt (copy editor); Michael Hamilton (proofreader); and Joanne Phillips (indexer). Daria Hupov (film editor, *Six Inches of Soil*) was wonderful, so promptly responding to our seemingly unceasing requests for extra raw footage, transcripts and film stills.

The book would not look as good as it does without the gracious permission of *Six Inches of Soil* animator Brook Morgan to adapt and expand on her graphics from the film to form the basis of the cover and internal layout and to designer Alex Lazarou for doing so so well. Finally, thanks to all the film participants who responded to check transcript texts and proofs.

Glossary

agroecology	application of ecological concepts and principals in farming
agroforestry	land-use management system in which combinations of trees are grown around or among crops or pasture
agronomist	expert in the science of soil management and crop production
agrosilvopastoral	land-use systems that combine a woody component (trees or shrubs) with livestock
alley coppice	combination of short rotation coppice in wide alleys with the production of high-value trees
alley cropping	agroforestry practice of planting rows of trees and/or shrubs to create alleys within which agricultural or horticultural crops are produced
anthropogenic carbon	release of carbon dioxide caused by human activities, which include the burning of fossil fuels, deforestation, land use and land-use changes, livestock production, fertilisation, waste management and industrial processes
arable	engaged in, produced by or being the cultivation of land with crops
Basic Payment Scheme (BPS)	until 2024 the main UK area-based farm subsidy, *see also* Sustainable Farming Incentive (SFI)
Belted Galloway	cattle; traditional Scottish breed of beef cattle, derived from the cattle of the Galloway region of south-western Scotland and established as a separate breed in 1921, well suited to living on poor upland pastures and in inclement weather
Big Ag	pejorative term for the large, often multinational, corporations that dominate the agri-food system
biodynamic	a holistic, ecological and ethical approach to farming, gardening, food and nutrition, based on concepts developed by Rudolph Steiner

black grass   widespread and aggressive weed that affects arable yields and causes contamination in harvests, *Alopecurus myosuroides*

break crop   a crop grown between main cereal cash crops in an arable rotation and taken to harvest (*compare* cover crop), they play important roles in fertility building, disease and weed control

brome   lowland weed common to south of England, *Bromus secalinus*

calcareous (soil)   mostly or partly composed of calcium carbonate; containing lime or being chalky

carbon sequestration   the capturing, removal and storage of carbon dioxide (CO_2) from the Earth's atmosphere in a carbon pool, such as the soil; part of the natural carbon cycle and is increasingly being recognised for its global warming mitigation potential

carcinogen   a substance, organism or agent capable of causing cancer

cash crop   a crop planted for the purpose of selling on the market or for export to make profit, as distinguished from subsistence crops planted for the purpose of self-supply of the farmer (e.g. livestock feed or personal food)

catch crop   *see* cover and break crops

CO_2e   carbon dioxide equivalent – the number of metric tonnes of CO_2 emissions with the same global warming potential as 1 metric tonne of another greenhouse gas (e.g. 1 kg of methane = 29.8 kg CO_2e)

combinable crops   crops harvested with a combine harvester, wheat, barley and oilseed rape are the most important combinable crops grown in the UK

Common Agricultural Policy (CAP)   the agricultural policy of the European Commission, introduced in 1962, it implements a system of agricultural subsidies (Single Farm Payment) and other programmes

companion cropping   a crop sown alongside the planned cash crop either before or at planting to help aid crop establishment by giving some protection from pests, increasing beneficial predatory insects and improving soil health; sometimes grown as a sacrificial crop

conservation-grazing	use of semi-feral or domesticated grazing livestock to maintain and increase the biodiversity of natural or semi-natural grasslands, heathlands, wood pasture, wetlands and many other habitats
controlled traffic farming	system using permanent wheel tracks (tramlines) to separate crop zones and traffic lanes, minimising the land area driven over; improves profitability and sustainability
conventional (industrial) agriculture	systems incorporating intensive use of agrochemicals, intensive tillage, monocropping and limited recycling of materials
coppice, coppicing	traditional method in woodland management of cutting a tree down to a stump, which in many species encourages new shoots to grow from the stump or roots, thus ultimately regrowing the tree, and allowing the harvesting of the cut timber
Corn Laws	the Corn Laws were tariffs and other trade restrictions on imported food and corn enforced in the UK in the 19th century (corn denoted all cereal grains, including wheat, oats and barley), designed to favour domestic producers over imports they ended up artificially keeping prices high, which benefitted landowners at the expense of the poor
council (county) farms	land usually owned and run by local councils that were set up at the end of the 19th century to provide a way into farming; many councils have sold them off to raise money; the number of farms and land area they cover are much reduced
Countryside Stewardship (CS) scheme	government financial incentives paid to farmers, foresters and land managers to look after and improve the environment, introduced in 1991, as of 2024 CS forms part of the new ELMs
cover crop	a crop grown between main cereal cash crops in an arable rotation and not taken to harvest (*compare* break crop), main role is to provide soil protection, they also play important roles in fertility building, disease and weed control, and provide forage for livestock

Devon (breed)   cattle; often referred to as Red Rubies due to their deep red-brown colour, Devon cattle are a native breed originating from the West Country; rising interest in low-input, environmentally sustainable production is driving a resurgence of the breed

drilling   mechanical means of creating furrows (openings) in the soil and metering seed in at a uniform rate: conventional drills work in tilled and partly tilled soil; direct drills work on undisturbed soil, placing the seed directly in the residues of the previous crop, *see also* no-till

enclosure   act of putting fences around land; refers to the appropriation of 'waste' or 'common land' and by doing so depriving commoners of their rights of access and privilege

the Enclosures   refers to 18th- and 19th-century acts of Parliament and the large scale of enclosure driven by the Agricultural Revolution

Environmental Land Management scheme (ELMs)   since 2024 the overarching UK government grant funding scheme for farmers, foresters and land management, based on the principle of public money for public goods and, unlike BPS, it will no longer subsidise farmers for food production, comprised of SFI, CS and Landscape Recovery

field-scale vegetable   vegetable grown on a field scale for human consumption, may be root vegetables such as carrots, potatoes or parsnips, or pulses such as peas or beans

finishing   feeding an energy-dense diet to livestock to promote rapid growth and muscle/meat gain and to optimise fat cover in preparation for slaughter

flerd   combination of the words herd and flock, simply refers to a group of multiple livestock species who are managed and graze as one group

foliar (feed)   direct application of liquid fertiliser to crop leaves; benefits compared to soil applications include quicker absorption, lower application volumes, less excess leaching

functional biodiversity	components of biodiversity that influence how an ecosystem operates or functions
fungal hyphae	long, branching, filamentous structure of a fungus, oomycete or actinobacterium; in most fungi, hyphae are the main mode of vegetative growth, and are collectively called a mycelium
ghost acres	displaced consumption of resources; historically, the Global North has outsourced production and the labour required into colonies (e.g. sugar plantations); currently mainly refers to the area of non-domestic land used to support consumption at home (e.g. soy grown in South America, often on deforested land, and imported to feed livestock)
green manure	fast-growing plants sown to cover bare soil, often used in vegetable production, their foliage smothers weeds and their roots prevent soil erosion; when dug into the ground while still green, they return valuable nutrients to the soil and improve soil structure, *see also* cover crop
Green Revolution	a period of technology transfer initiatives that saw greatly increased crop yields; these changes in agriculture began in developed countries in the early 20th century and spread globally till the late 1980s, also known as the Third Agricultural Revolution
greenwashing	deliberately claiming or creating the impression that a company or other entity is doing more to protect the environment than it is; greenwashing promotes false solutions to the climate crisis that distract from and delay concrete and credible action (*see* Interlude II)
guano	accumulated excrement and remains of birds, bats and seals, valued as fertiliser
Haber–Bosch process	primary method in synthesising ammonia from nitrogen and hydrogen (see Chapter 1)
headland	outer edges of the field where farm machinery turns during field operations such as drilling and ploughing (also known as a turnrow)
heifer	young female cattle that have not yet borne calves

herbal ley	grasslands or additional seeding added to them, made up of legume, herb and grass species, both temporary and permanent; leys easily fit into arable and mixed farming rotations
horticulture	agriculture concerned with growing plants that are used by people for food, for medicinal purposes and for ornamentation
hunger gap	the period, usually in April, May and early June, after the winter crops have ended but before the new season's plantings are ready to harvest
industrial agriculture	*see* conventional
intercropping	growing two or more crops alongside one another with the goal of increasing the yield per area of land by making better use of the resources in the soil than would otherwise be used by a single crop; additionally the crops can be mutually supportive; similar to but differs from companion cropping in that all crops are grown for harvest
Knepp	refers to the Knepp Estate, Sussex, a pioneer rewilding project
land-take	use of undeveloped land (agricultural and natural) for human settlements and transportation infrastructure
legume	plant in the family Fabaceae, or the fruit or seed of such a plant; when dry, the seed is also called a pulse; legumes are grown primarily for human consumption, for livestock forage and silage and as soil-enhancing green manure; many are nitrogen fixing
lodge	permanent displacement of a stem (or part of a stem) from a vertical posture; often caused by high wind speeds, made worse by wet conditions; in extreme situations, crops may be forced flat
mob grazing	short-duration, high-density grazing with a longer than usual grass recovery period
mucilage	thick gluey substance produced by nearly all plants and some microorganisms; in plants this binds soil particles

mulch
material applied to the surface of soil; applied to conserve of soil moisture, improve soil fertility and health, reduce weed growth and enhance visual appeal

mycorrhizae
symbiotic associations between a fungus and a plant; refers to the role of the fungus in the plant's rhizosphere (its root system); mycorrhizae play important roles in plant nutrition, soil biology and soil chemistry

no-till
agricultural technique for growing crops or pasture without disturbing the soil through tillage; no-till farming decreases the amount of soil erosion tillage can cause; also known as zero tillage or direct drilling

NPK
nitrogen (N), phosphorus (P) and potassium (K)

organophosphates
organic compounds containing phosphorus; primary components in herbicides, pesticides and insecticides

overseed
to plant seeds directly in an existing plant ground cover, most often grass

overstory
top foliage from multiple trees that combine to create an overhang or canopy in a wooded area or forest

Oxford Sandy and Black
pig; breed of domestic pig originating in Oxfordshire; named for its colour, which is a base of sandy brown with black patches; sometimes called the 'Plum Pudding' or Oxford Forest pig

pan (soil)
a dense layer of soil, usually found below the uppermost topsoil layer; there are different types of pan, all sharing the general characteristic of being a distinct soil layer that is largely impervious to water; some pans are formed by deposits in the soil that fuse and bind the soil particles (such as dissolved silica or matrices formed from iron oxides and calcium carbonate), others are anthropogenic, such those formed by compaction from repeated ploughing or by heavy traffic; also known as a hardpan

pannage
practice of releasing livestock, most often pigs, in a forest to feed on fallen acorns, beechmast, chestnuts or other nuts; historically, it was a right or privilege granted to local people on common land or in royal forests across much of Europe; also known as 'common of mast'

pasture chicken   bird allowed to roam free in a pasture, spends all of its time on grass, out in fresh air; partially self-feeding eating seeds, insects and earthworms found in the field

permanent pasture   land used permanently (for several consecutive years, normally 5 years or more) to grow herbaceous fodder, forage or energy purpose crops, through cultivation (sown) or naturally (self-seeded), and which is not included in the crop rotation on the holding

perennial   in botany, a perennial plant is one that lives more than 2 years; the term is often used to differentiate a plant from shorter-lived annuals and biennials

peri-urban   (peripheral – around, about or beyond) non-urban landscapes adjacent to or surrounding metropolitan settlements; a peri-urban area can be defined in relation to a nearby metropolitan area on its inner boundary, a rural area on its outer boundary or as the land in between

poached, poaching   damage caused to turf or sward by the feet of livestock; hooves cause compaction of the soil surface, leaving depressions that can be 10–12 cm deep, which can form an almost continuous layer of grey anaerobic soil in which natural activity, carried out by soil micro-organisms, is low; most often caused by over-stocking or animals dwelling in gateways or at feed stations; poached land in turn suffers from soil erosion, compaction and waterlogging

public goods   in economics a product or service that every member of a society can consume freely without reducing its availability to others; in relation to agriculture it encompasses the safeguarding and regeneration of natural capital and provision of public goods such as beauty, heritage and engagement; farmland and woodland can provide public goods in the forms of thriving plants and wildlife as well as contributing to the provision of cleaner air and water, see also ELMs

ramial (woodchip)   un-composted woodchip made from smaller diameter (≤7 cm) younger tree branches; nutritionally these are the richest parts of trees, with young tree branches containing as much as 75% of the minerals, amino acids, proteins, phytohormones and enzymes found in the tree

rewilding   ecological restoration aimed at increasing biodiversity and restoring natural processes; differs from other forms of ecological restoration in that rewilding aspires to reduce human influence on ecosystems

rhizome   a modified subterranean plant stem that sends out roots and shoots from its nodes; rhizomes are also called creeping rootstalks or just rootstalks; rhizomes develop from axillary buds and grow horizontally and retain the ability to allow new shoots to grow upwards

riparian   relating to or situated on the banks of a river; in ecology specifically relating to wetlands adjacent to rivers and streams

roller crimping   practice in which the plant, often a cover crop, is flattened to the soil surface and its stems 'crimped'; the crimping action helps crush the walls of the stem to facilitate drying down of the plant

rogueing   act of identifying and removing plants with undesirable characteristics from agricultural fields; rogues are removed from the fields to preserve the quality of the crop being grown; plants removed may be diseased, be of an unwanted variety or undesirable for other reasons

Romney (breed)   sheep; formerly called the Romney Marsh or locally the Kent, is a breed of sheep originating from the marshy area of that country, a dual purpose (fleece and meat) breed, that was exported with colonialism and is the predominate breed in New Zealand

ruminant   herbivorous grazing or browsing mammals that are able to acquire nutrients from plant-based food by fermenting it in a specialised stomach (rumen) prior to digestion, principally through microbial actions; comprise cattle, sheep, antelopes, deer, giraffes and their relatives

seedbank natural storage of seeds, often dormant, within the soil of most ecosystems

shelterbelts a windbreak of planting usually made up of one or more rows of trees or shrubs planted in such a manner as to provide shelter from the wind and to protect soil from erosion

silvoarable crops grown simultaneously with a long-term tree crop; trees are grown in rows with wide alleys in-between for cultivating crops; crops provide annual income while the tree crop matures; from Latin *silva* 'wood'

silvopastoral trees deliberately introduced into a livestock forage production system (or, less commonly, forage introduced into a tree production system), the whole designed to produce a high-value tree component, while continuing to produce forage and livestock component indefinitely or for a significant time; from Latin *silva* 'wood'

Single Farm Payment (SFP) main EU agricultural subsidy, introduced in 2003 it pays farmers on a per-hectare basis; until the UK left the EU this was administered by the UK government and was paid out as the Basic Farm Payment

soft (& hard) wheat soft wheat flour is lower in gluten and high in starch; hard wheat generally has a higher gluten (and protein) content; typically, bread flour is made entirely from hard wheat, cake flour is made entirely from soft wheat, and most other baked goods use a combination of flours

soil organic carbon a measurable component of soil organic matter; organic matter makes up 2–10% of most soil's mass and has an important role in the physical, chemical and biological function of agricultural soils; organic matter contributes to nutrient retention and turnover, soil structure, moisture retention and availability, degradation of pollutants and carbon sequestration

stacking layering of crops or enterprises on a farm, *see* Interlude VI

stand (forestry)	a contiguous community of trees sufficiently uniform in composition, structure, age, size, class, distribution, spatial arrangement, condition or location on a site of uniform quality to distinguish it from adjacent communities
store cattle	animals c.18 months old 'stored' over winter before being finished (fattened) prior to slaughter; *see also* finishing
subsidies	a benefit given to an individual, business or institution, usually by the government; can be direct (such as cash payments) or indirect (such as tax breaks); *see* BPS, CAP, CS, SFI and SFP
subsoiler (machine)	a tillage tool, angled wings are used to lift and shatter the (hard)pan that builds up due to compaction; the design provides deep tillage, loosening soil deeper than a tiller or plough is capable of reaching
subsoiling	the operation (using a subsoiler) to break up pans or compacted layers and loosen the subsoil with minimal disruption of the surface
Sustainable Farming Incentive (SFI)	introduced in 2024, the basic component of ELMs, SFI financially rewards farmers for employing environmentally sustainable land management practices, *see also* ELMs
sward	surface layer of ground containing a mat of grass and grass roots
swath	a row or line of grass, corn or other crop as it falls or lies when mown or reaped
swathing	the process of cutting a crop so it lies in a swath to dry and be collected
systemic herbicide	herbicides that are absorbed and transported through the plant's vascular system, killing the entire plant; in contrast to contact herbicides that kill the part of the plant in contact with the chemical leaving the roots, which may survive and the plant may regrow
Tamworth (breed)	pig; a very hardy animal, suited to an outdoor system; ginger coat protects the breed from sunburn

understory	in forestry and ecology, also known as underbrush or undergrowth, includes plant life growing beneath the forest canopy (overstory) without penetrating it to any great extent, but above the forest floor; in agriculture may refer to plants forming a lower level of growth, such as companion clover plants under a main cereal crop
universal basic income	social welfare proposal in which all citizens of a given population regularly receive a minimum income in the form of an unconditional transfer payment, i.e. without a means test or need to work
Welsh Black	cattle; a dual-purpose breed of cattle native to Wales; hardy breed providing high quality meat and milk
Welsh Mountain	sheep; small, hardy sheep from the higher parts of the Welsh mountains
windrow	*see* swath
wood pasture	most often historic areas of silvopasture (former medieval hunting forests or wooded commons), though some are more recent (those on large country estates); commonly managed through grazing; ranging from mixtures scrub and denser woodland groves, to more open grassland or heathland with scattered trees

Notes

CHAPTER 1

IT'S A MIRACLE, BUT THAT MIRACLE HAS CREATED A DISASTER

1 Dimbleby, Henry, *The National Food Strategy: The Plan*, https://www.nationalfoodstrategy.org.

2 Food Supplies Vol. 453, c.898: debated on Monday 12 July 1948.

3 Just 3 of the 2045 examples returned by a search of the https://www.britishnewspaperarchive.co.uk with the keywords 'food', 'population', famine' for the years 1940–49. *Western Times*, Friday 29 August 1947 © Reach PLC, reproduced with permission; *Western Daily Press*, Wednesday 15 May 1946 © Reach PLC, reproduced with permission; *Dundee Courier and Advertiser*, Wednesday 3 September 1947 © D.C. Thomson & Co. Ltd, produced with permission.

4 Global Food Security, Food security – a history, https://www.foodsecurity.ac.uk/challenge/food-security-history/.

5 Agriculture Bill Vol. 432, c. 835: debated on Tuesday 28 January 1947.

6 Victor M. Shorrocks, *Conventional and Organic Farming: A Comprehensive Review through the Lens of Agricultural Science* (5m Books 2017), p. 18, see chapter 2 for an introduction to the history of fertilisers.

7 Shorrocks, *Conventional and Organic Farming*, p. 22.

8 For a detailed and scathing assessment of effects of subsides on the UK environment see Graham Harvey's *The Killing of the Countryside* (Vintage 1998).

9 Dimbleby, *The National Food Strategy*, p. 40.

10 The Food Foundation's The Broken Plate 2023 reports that 'healthy foods are over twice as expensive per calorie as less healthy foods' and that 'Nearly 9,600 diabetes-related amputations are carried out on average per year – an increase of 19% in six years.' https://www.foodfoundation.org.uk/sites/default/files/2023-10/TFF_The%20Broken%20Plate%202023_Digital_FINAL..pdf.

11 M. Berners-Lee, C. Kennelly, R. Watson & C.N. Hewitt, Current global food production is sufficient to meet human nutritional needs in 2050 provided there is radical societal adaptation. *Elementa: Science of the Anthropocene* 1 January 2018; 6 52. https://doi.org/10.1525/elementa.310.

12 A.-M. Berenice Mayer, L. Trenchard & F. Rayns, Historical changes in the mineral content of fruit and vegetables in the UK from 1940 to 2019: a concern for human nutrition and agriculture. *International Journal of Food Sciences and Nutrition* 2022; 73:3, 315-326. https://doi.org/10.1080/09637486.2021.1981831.

13 Dimbleby, *The National Food Strategy*, pp. 89–90 & Figure 9.3.

CHAPTER 2

REGENERATIVE FARMING, WHAT IS IT?

1 https://www.biodynamics.com/steiner.html.

2 Marina O'Connell, *Designing Regenerative Food Systems* (Hawthorn Press, 2022), p. 28.

3 As written about by F.H. King, in his book *Farmers of Forty Centuries:*

Organic Farming in China, Korea, and Japan (Dover, 2004).

4   www.soilassociation.org/who-we-are/organic-principles.

5   Bill Mollison, *Permaculture: A Designers' Manual* (Tagari, 1988).

6   Soil Association, *The Agroforestry Handbook: Agroforestry for the UK*, https://www.soilassociation.org/media/19141/the-agroforestry-handbook.pdf, p.11.

7   'The Peasants' Way', https://viacampesina.org/en/.

8   La Via Campesina represents 182 member organisations, in 81 countries encompassing over 200 million peasants, https://viacampesina.org/en/what-are-we-fighting-for/. Peasant, often used as a derogatory term, has been deliberately chosen and reclaimed by the agroecological movement to mean 'people of the land'. The Landworkers' Alliance is the UK's La Via Campesina member organisation, https://landworkersalliance.org.uk/our-vision/.

9   https://viacampesina.org/en/declaration-of-the-international-forum-for-agroecology/.

10 FAO, The 10 Elements of Agroecology: Guiding the transition to sustainable food and agricultural systems, http://www.fao.org/3/I9037EN/i9037en.pdf.

11 HLPE, Agroecological and other innovative approaches for sustainable agriculture and food systems that enhance food security and nutrition. A report by the High Level Panel of Experts on Food Security and Nutrition of the Committee on World Food Security, Rome, https://www.fao.org/3/ca5602en/ca5602en.pdf, p.41.

12 Professor John Crawford of Rothamsted Research has pointed out that, at the current rate of degradation, we only have enough soil left globally to provide food for another 60 years (https://sustainablefoodtrust.org/

news-views/ten-things-about-soil.

13 See Victor M. Shorrocks, *Conventional and Organic Farming: A Comprehensive Review through the Lens of Agricultural Science* (5m Books 2017), ch. 2, for a concise yet comprehensive summary of early agricultural developments and the key people involved.

14 Summaries of the five principles collated from several sources including the RASE Farm of the Future report, https://www.rase.org.uk/content/large/documents/reports/farm_of_the_future-_journey_to_net_zero.pdf; https://groundswellag.com/principles-of-regenerative-agriculture/; https://agricaptureco2.eu/the-six-principles-of-regenerative-farming-why-are-they-important/.

INTERLUDE 1

ARE WE USING OUR FARMED LAND WISELY?

1   https://stateofnature.org.uk.

2   RSPB, Making our land work for nature, climate and people, https://community.rspb.org.uk/ourwork/b/nature-s-advocates/posts/making-our-land-work-for-nature-climate-and-people.

3   https://www.sciencedirect.com/science/article/pii/S259033222300444X.

4   RSPB, Sparing or sharing?, https://community.rspb.org.uk/ourwork/b/science/posts/sparing-or-sharing; Fred Lewsey, Concentrate farming to leave room for species and carbon, https://www.cam.ac.uk/stories/landsparing; *The National Food Strategy: The Plan*, https://www.nationalfoodstrategy.org.

5   FAO, The 10 elements of agroecology, https://www.fao.org/3/i9037en/i9037en.pdf.

6   Sustain, Regenerative farming under the spotlight, https://www.sustainweb.org/blogs/oct21-is-regenerative-farming-legitimate/.

7 George Monbiot, *Regenesis: Feeding the World without Devouring the Planet* (Allen Lane 2022); Fermenting a revolution, https://www.monbiot.com/2022/11/26/fermenting-a-revolution/.

8 https://landworkersalliance.org.uk/wp-content/uploads/2018/10/Response-to-George-Monbiot-livestock-critique.pdf.

9 For what this means in practice see Sustain, Is there any other sector like farming?, https://www.sustainweb.org/blogs/mar22-is-any-sector-like-farming/, and FFCC, Multifunctional Land Use Framework', https://ffcc.co.uk/multifunctional-land-use-framework.

CHAPTER 3
THE SCIENCE OF SOIL

1 Sheldrake, Merlin, *Entangled Life : How Fungi Make Our Worlds, Change Our Minds and Shape Our Futures* (Bodley Head, 2020).

2 FAO, State of Knowledge of Soil Biodiversity, https://www.fao.org/3/cb1929en/cb1929en.pdf.

3 Graves et al., The total costs of soil degradation in England and Wales. *Ecological Economics*, 119 (2015), 399–413, https://doi.org/10.1016/j.ecolecon.2015.07.026.

4 Environment Agency, The state of the environment: soil, https://assets.publishing.service.gov.uk/media/5cf4cbaf40f0b63affb6aa55/State_of_the_environment_soil_report.pdf.

5 Masters, Nicole, *For the Love of Soil: Strategies to Regenerate Our Food Production Systems* (Bowker, 2019).

6 https://sustainablesoils.org/soil-carbon-code/minimum-requirements.

7 https://www.rothamsted.ac.uk/news/

potential-soils-sequester-carbon-being-seriously-overestimated.

8 Masters, *For the Love of Soil.*

9 AHDB Soil biology & health partnership, https://ahdb.org.uk/soil-biology-and-soil-health-partnership; Soil health scorecard, https://ahdb.org.uk/knowledge-library/the-soil-health-scorecard.

10 EFRA, Soil health, https://committees.parliament.uk/committee/52/environment-food-and-rural-affairs-committee/news/198809/take-soil-as-seriously-as-air-and-water-to-protect-food-supply-and-environment-mps-urge/.

INTERLUDE II
REGULATION, GREENWASHING AND CO-OPTION

1 Table, What is regenerative agriculture?, https://www.tabledebates.org/building-blocks/what-is-regenerative-agriculture.

2 Groundswell, 5 Principles of Regenerative Agriculture, https://groundswellag.com/principles-of-regenerative-agriculture/.

3 Wildfarmed, Farming standards, https://www.wildfarmed.co.uk/products/wildfarmed-regenerative-standards.

4 X. Li, J. Storkey, A. Mead et al., A new Rothamsted long-term field experiment for the twenty-first century: principles and practice. *Agronomy for Sustainable Development* 43, 60 (2023), https://doi.org/10.1007/s13593-023-00914-8.

5 Z. Luo, E. Wang & O.J. Sun, Can no-tillage stimulate carbon sequestration in agricultural soils? A meta-analysis of paired experiments. *Agriculture, Ecosystems & Environment* 139(1) (2010), 224–231, https://doi.org/https://doi.org/10.1016/j.agee.2010.08.006.

6   https://www.pepsico.com/who-we-are/
    our-commitments/pepsico-positive.

7   pepsico-regenerative-agriculture-
    scheme-rules.pdf

8   Marion Nestle, PepsiCo' push
    into regenerative agriculture:
    real or greenwashing?, https://
    www.foodpolitics.com/2022/09/
    pepsico-push-into-regenerative-
    agriculture-real-or-greenwashing/.

9   https://www.fairr.org/resources/reports/
    regenerative-agriculture-four-labours.

10 Grantham Research Institute,
    Corruption and integrity risks in
    climate solutions: an emerging global
    challenge, https://www.lse.ac.uk/
    granthaminstitute/publication/
    corruption-and-integrity-risks-in-
    climate-solutions; Food Navigator,
    Food giants at risk of 'greenwashing',
    https://www.foodnavigator.com/
    Article/2023/01/31/Food-giants-at-
    risk-of-greenwashing-over-regenerative-
    agricultural-practices-report-warns;
    Provenance, 5 food and drink
    brands called out for greenwashing,
    https://www.provenance.org/
    news-insights/5-food-and-drink-
    brands-called-out-for-greenwashing-
    and-the-lessons-we-can-learn.

11 Khangura Ravjit, David Ferris,
    Cameron Wagg & Jamie Bowyer,
    Regenerative agriculture—a
    literature review on the practices
    and mechanisms used to improve
    soil health. *Sustainability* 15, no. 3
    (2023): 2338, https://doi.org/10.3390/
    su15032338.

CHAPTER 4

REGENERATIVE MIXED FARM, THE PINK PIG, SCUNTHORPE, LINCOLNSHIRE

1   In January 2024 Riverford's Get Fair
    About Farming campaign, https://
    getfairaboutfarming.co.uk/#charter,
    secured enough petition signatures to
    force government debate.

2   C.M. Benbrook, How did the US
    EPA and IARC reach diametrically
    opposed conclusions on the
    genotoxicity of glyphosate-based
    herbicides? *Environmental Science
    Europe* 31, 2 (2019). https://doi.
    org/10.1186/s12302-018-0184-7.
    See among others M. Kogevinas,
    Probable carcinogenicity of
    glyphosate. *BMJ* 2019; 365, https://
    doi.org/10.1136/bmj.l1613 and J.V.
    Tarazona, D. Court-Marques, M.
    Tiramani et al., Glyphosate toxicity
    and carcinogenicity: a review of the
    scientific basis of the European Union
    assessment and its differences with
    IARC. *Arch Toxicol.* 91, 8 (2017),
    2723–2743, https://doi.org/10.1007%
    2Fs00204-017-1962-5.

3   https://www.unep.org/
    news-and-stories/story/
    four-reasons-why-world-needs-limit-
    nitrogen-pollution.

4   https://sustainablesoils.org/about-
    soils/soils-and-public-health; https://
    www.medicalnewstoday.com/
    articles/66840#1; C. A. Lowry et
    al., Identification of an immune-
    responsive mesolimbocortical
    serotonergic system: Potential
    role in regulation of emotional
    behavior. *Neuroscience* 146, 2 (2007),
    756–772, https://doi.org/10.1016/j.
    neuroscience.2007.01.067.

INTERLUDE III

SUBSIDIES, BREXIT AND TRADE

1   Sustain, We have an Agriculture
    Act – but let's not relax now,
    https://www.sustainweb.org/blogs/
    nov20-new-agriculture-act2020/.

2   IDDRI, An agroecological Europe
    in 2050: multifunctional agriculture
    for healthy eating, https://www.iddri.
    org/en/publications-and-events/
    study/agroecological-europe-2050-

multifunctional-agriculture-healthy-eating; Modelling an agroecological UK in 2050 – findings from TYFA-REGIO, https://www.iddri.org/en/publications-and-events/study/modelling-agroecological-uk-2050-findings-tyfa-regio.

3 Sustain, UK public wants protection for high food standards and the environment in future trade deals, https://www.sustainweb.org/news/nov20-which-trade-conversation-food-environment/.

4 Sustain, Farmers call on MPs to support food and farming standards in the Ag Bill, https://www.sustainweb.org/news/nov20-farmers-letter-mps/.

5 Pesticide Action Network UK, Toxic trade, https://www.pan-uk.org/toxic-trade/; Sustain, Good Food Trade Campaign, https://www.sustainweb.org/news/nov20-antibiotics-future-trade-deals-us-australia-canada-nz/.

6 Sustain, Trick or trade: the impacts of free trade agreements on food environments and child obesity, https://www.sustainweb.org/reports/trick-or-trade-report/.

CHAPTER 5
ORGANIC MARKET GARDEN, SWEETPEA, CAXTON, CAMBRIDGESHIRE

1 World Wide Opportunities on Organic Farms, https://wwoof.net.

2 The two leading certification bodies, among others, in England are OF&G (see page 278) and the Soil Association (see page 279). The Organic Research Centre's *Organic Farm Management Handbook 2023* (Organic Farmers & Growers, 2023) is a good source of additional advice.

3 https://communitysupported agriculture.org.uk.

4 https://cambridgefoodhub.org.

5 Jules Pretty, *The Low-Carbon Good Life* (Taylor & Francis, 2022) and *Thirty for 30: Cutting your Carbon*. Essex Rural Partnership, https://admin.essexruralpartnership.org.uk/public/uploads/all/TsxkCe07piRrbtybbuVZ4P1zhhLRfJJru7AUWL2Y.pdf.

6 Tim Lang, *Feeding Britain: Our Food Problems and How to Fix Them* (Pelican, 2020). https://ahdb.org.uk/news/consumer-insight-how-much-money-are-uk-consumers-spending-on-food-since-the-cost-of-living-crisis.

7 https://growingcommunities.org.

8 Marina O'Connell, *Designing Regenerative Food Systems and Why We Need Them Now* (Hawthorn Press, 2022).

9 https://rodaleinstitute.org/education/resources/regenerative-agriculture-and-the-soil-carbon-solution.

10 https://basicincome.org/news/2023/12/ubi-for-farmers-campaign-launch-report-insights-and-discussion.

11 https://landworkersalliance.org.uk/lwa-wlcomes-new-sfi-commitment-but-feelsl-ambition-still-lagging. However, page 101 of the SFI 2023 handbook v4 January 2024 maintains the 5 ha criteria as it permits SFI applications only from previously eligible BPS farms. It goes on to vaguely say 'In future, we'll allow a wider range of farmers to apply. This will not happen before 2024.'

12 https://www.worldwildlife.org/stories/what-is-the-sixth-mass-extinction-and-what-can-we-do-about-it.

INTERLUDE IV
FOOD SECURITY AND UK SELF-SUFFICIENCY

1 FAO, Rome Declaration on World Food Security, https://www.fao.org/3/w3613e/w3613e00.htm.

2 ECIU, Storms, floods, droughts and an unstable food supply, https://eciu.net/

insights/2023/storms-floods-droughts-and-an-unstable-food-supply.

3 https://assets.publishing.service.gov.uk/media/62874ba08fa8f55622a9c8c6/United_Kingdom_Food_Security_Report_2021_19may2022.pdf.

4 IPBES-Food, A Long Food Movement: Transforming Food Systems by 2045, http://www.ipes-food.org/pages/LongFoodMovement.

5 F. Rauber et al., Ultra-processed food consumption and indicators of obesity in the United Kingdom population (2008-2016). *PLoS ONE* 15, 5, e0232676, https://doi.org/10.1371/journal.pone.0232676.

6 IDDRI, An agroecological Europe in 2050: multifunctional agriculture for healthy eating, https://www.iddri.org/en/publications-and-events/study/agroecological-europe-2050-multifunctional-agriculture-healthy-eating.

7 See research and definitions at https://www.eating-better.org/.

8 See Landworkers' Alliance, Horticulture Across Four Nations, https://landworkersalliance.org.uk/new-report-horticulture-across-four-nations-jan-2024/#:~:text=Our%20brand%20new%20report%2C%20"Horticulture,on%20sourcing%20fresh%2C%20agroecological%20produce; Sustain, Fringe farming: peri-urban agroecology towards resilient food economies and public goods, https://www.sustainweb.org/publications/feb22-fringe-farming/.

9 Sustain, Beyond the farmgate: unlocking the path to farmer-focused supply chains and climate-friendly, agroecological food systems, https://www.sustainweb.org/publications/beyond-the-farmgate/.

CHAPTER 6
PASTURE-FED BEEF, TREVEDDOE, BODMIN, CORNWALL

1 Carbon Calling ran again successfully over two sites in September 2023 and is planned to return in 2024. https://www.carboncalling.farm.

INTERLUDE V
APPLYING AGROFORESTRY IN MIXED REGENERATIVE FARMING

1 Soil Association, *The Agroforestry Handbook: Agroforestry for the UK*, https://www.soilassociation.org/media/19141/the-agroforestry-handbook.pdf.

2 https://www.agroforestryshow.com. Videos of recorded sessions are available on the show's website.

3 *The Agroforestry Handbook*, p. 51.

CHAPTER 7
SOIL AS SOUL

1 In interview for '6 Inches' at ORFC, January 2023.

2 George Monbiot, *Regenesis: Feeding the World without Devouring the Planet* (Allen Lane 2022).

3 Chris Smaje, *Saying No to a Farm Free Future* (Chelsea Green, 2023).

4 Jake Fiennes, *Land Healer* (2022, Witness Books).

5 Anne Biklé and David R. Montgomery, *What Your Food Ate: How to Heal Our Land and Reclaim Our Health* (W. W. Norton, 2022).

6 https://www.peoplesfoodpolicy.org.

7 Satish Kumar, *Soil Soul Society: A New Trinity for Our Time* (Leaping Hare Press, 2017).

8 Scott Chaskey, *Soil and Soul: Cultivation and Kinship in the Web of Life* (Milkweed, 2023).

9   Sophie Strand, *The Flowering Wand: Rewilding the Sacred Masculine* (Inner Traditions, 2022).

INTERLUDE VI
ENTERPRISE STACKING

1   Tim's report, https://www. nuffieldscholar.org/reports/gb/2011/ understanding-and-implementing-sustainability. The Nuffield Farming Scholarships Trust is a registered charity whose aim is to inspire passion in people and develop their potential to lead positive change in farming and food. They award life-changing scholarships that unlock individual potential and broaden horizons through study and travel overseas, with a view to developing the farming and agricultural industries.

CHAPTER 8
HOW DO WE BUILD A SUSTAINABLE FUTURE FOR FOOD AND FARMING?

1   Check out https://farmsunday.org and https://agroforestryopenweekend.org, and better still take the opportunity to talk to a local farmer or grower.
2   L. Michaels & M. Puig de la Bellacasa, 'It's only really when I put my hands in the soil that I feel at home'. Soil care and ecological belonging in urban growers' practices. *Environmental Humanities* (forthcoming, 2024); L. Michaels, 'The soils of Leicester are rich, but it's what they give ... not just the vegetables'. Ecological belongings: voices of Leicester growers. Zine and exhibition in Leicester during 2024 (with illustrations by Nico Vass). (forthcoming, 2024). Project website: Ecological Belongings: Transforming Soil Cultures Through Science, Art, and Activism accessible at: https:// warwick.ac.uk/fac/cross_fac/cim/ research/ecological-belongings/.
3   Sustain, Unpicking food prices: Where does your food pound go, and why do farmers get so little?, https://www.sustainweb.org/reports/ dec22-unpicking-food-prices/.

Further reading

A list of the books and major reports mentioned in the text along with suggested reading from our farmers and experts. It you are looking for somewhere to start, we suggest those in the top section.

Brown, Gabe, *Dirt to Soil: One Family's Journey into Regenerative Agriculture* (Chelsea Green, 2018)

Langford, Sarah, *Rooted: How regenerative farming can change the world* (Penguin, 2023)

Massey, Charles, *Call of the Reed Warbler: a New Agriculture, a New Earth* (Chelsea Green, 2017)

Montgomery, David, *Dirt: The Erosion of Civilizations* (University of California Press, 2007)

O'Connell, Marina, *Designing Regenerative Food Systems and Why We Need Them Now* (Hawthorn Press, 2022)

Rebanks, James, *English Pastoral: An Inheritance* (Penguin, 2021)

Wall Kimmerer, Robin, *Braiding Sweetgrass: Indigenous Wisdom, Scientific Knowledge, and the Teachings of Plants* (Milkweed, 2020)

Balfour, Eve, *The Living Soil* (Faber & Faber, 1943)

Berners-Lee, Mike, *The Carbon Footprint of Everything* (Greystone Books, 2022)

Berners-Lee, Mike, *There Is No Planet B: A Handbook for the Make or Break Years* (Cambridge University Press, 2021)

Biklé, Anne and Montgomery, David R., *What Your Food Ate: How to Heal our Land and Reclaim our Health* (W. W. Norton, 2022)

Blair, Robert, *A Practical Guide to the Feeding of Organic Farm Animals: Pigs, Poultry, Cattle, Sheep and Goats* (5m Books, 2017)

Blowey, Roger, W., *Cattle Lameness and Hoofcare 3rd Edition* (5m Books, 2015)

Blowey, Roger, W., *The Veterinary Book for Dairy Farmers 4th Edition* (5m Books, 2016)

Carson, Rachel, *Silent Spring* (Houghton Mifflin, 1962)

Chaskey, Scott, *Soil and Soul: Cultivation and Kinship in the Web of Life* (Milkweed, 2023)

Coleman, Elliot, *New Organic Grower: a Master's Manual of Tools and Techniques for the Home and Market Gardener* (Chelsea Green, 2018)

Collyns, Kate, *Gardening for Profit: from Home Plot to Market Garden* (UIT/ Green Books, 2013)

Conford, Philip, *The Origins of the Organic Movement* (Floris, 2001)

Dawling, Pam, *Sustainable Market Farming: Intensive Vegetable Production on a Few Acres* (New Society Publishers, 2013)

Deppe, Carol, *The Resilient Gardener: Food Production and Self-reliance in Uncertain Times* (Chelsea Green, 2014)

Dimbleby, Henry, *Ravenous: How To Get Ourselves And Our Planet Into Shape* (Profile Books, 2023)

Dimbleby, Henry, *The National Food Strategy: The Plan*, https://www.nationalfoodstrategy.org

Dunn, Peter, *The Goatkeeper's Veterinary Book 4th Edition* (5m Books, 2007)

Fiennes, Jake, *Land Healer* (2022, Witness Books)

Gerrish, Jim, *Keeping It Green: A Handbook for Creating & Managing Irrigated Pasture* (Chelsea Green, 2023)

Gerrish, Jim, *Kick the Hay Habit: A Practical Guide to Year-Around Grazing* (Chelsea Green, 2010)

Gerrish, Jim, *Management-intensive Grazing: The Grassroots of Grass Farming* (Chelsea Green, 2004)

Giles, Michaela, *The Commuter Pig Keeper: A Comprehensive Guide to Keeping Pigs when Time is your Most Precious Commodity* (5m Books, 2006)

Haraway, D. J., *Staying with the Trouble* (Duke University Press, 2016)

Hartman, Ben, *The Lean Farm: How to Minimize Waste, Increase Efficiency, and Maximize Value and Profits with Less* (Chelsea Green, 2015)

Harvey, Graham, *The Killing of the Countryside* (Vintage, 1998)

Haslett-Marroquin, Reginaldo, *In the Shadow of Green Man: My Journey from Poverty and Hunger to Food Security and Hope* (Acres, 2017)

Henderson, David C., *The Veterinary Book for Sheep Farmers* (5m Books, 2002)

Hervé-Gruyer, Perrine & Hervé-Gruyer, Charles, *Miraculous Abundance: One Quarter Acre, Two French Farmers, and Enough Food to Feed the World* (Chelsea Green, 2016)

Hird, Vicki, *Perfectly Safe to Eat? The Facts on Food* (Women's Press, 2001)

Hird, Vicki, *Rebugging the Planet: The Remarkable Things that Insects (and Other Invertebrates) Do – And Why We Need to Love Them More* (Chelsea Green Publishing, 2021)

HRH The Prince of Wales, *Prince's Speech: On the Future of Food* (Rodale, 2012)

IDDRI, 'An agroecological Europe in 2050: multifunctional agriculture for healthy eating', https://www.iddri.org/en/publications-and-events/study/agroecological-europe-2050-multifunctional-agriculture-healthy-eating

IDDRI, 'Modelling an agroecological UK in 2050 – findings from TYFA-REGIO', https://www.iddri.org/en/publications-and-events/study/modelling-agroecological-uk-2050-findings-tyfa-regio.

Kay, Niva & Kay, Yotam, *The Abundant Garden: a Practical Guide to Growing a Regenerative Home Garden* (Allen & Unwin, 2022)

King, F.H., *Farmers of Forty Centuries: Organic Farming in China, Korea, and Japan* (Dover, 2004)

Kumar, Satish, *Soil Soul Society: A New Trinity for Our Time* (Leaping Hare Press, 2017)

Landworkers' Alliance, *With the Land: Reflections on Land Work and 10 Years of the LWA* (LWA, 2023)

Landzettel, Marianne, *Regenerative Agriculture: Farming with Benefits. Profitable Farms. Healthy Food. Greener Planet* (Martin Kunz, 2021)

Lang, Tim and Heasman, Michael, *Food Wars: The Global Battle for Mouths, Minds and Markets* (Routledge, 2015)

Lang, Tim, Barling, David, and Caraher, Martin, *Food Policy: Integrating Health, Environment And Society* (Oxford University Press, 2009)

Lang, Tim, *Feeding Britain: Our Food Problems and How to Fix Them* (Pelican, 2020)

Lewis, Kahty, Tzilivakis, John, Warner, Doug and Green, Andy, *Agri-environmental Management in Europe: Sustainable Challenges and Solutions – From Policy Interventions to Practical Farm Management* (5m Books, 2018)

Mahendran, Sophie (ed.), *Handbook of Calf Health and Management* (5m Books, 2021)

Masters, Nicole, *For the Love of Soil: Strategies to Regenerate Our Food Production Systems* (Bowker, 2019)

Mills, Olivia, *Practical Sheep Dairying: Care and Milking of the Dairy Ewe* (Thorsons, 1989)

Mollison, Bill, *Permaculture: A Designers' Manual* (Tagari, 1988)

Monbiot, George, *Regenesis: Feeding the World without Devouring the Planet* (Allen Lane, 2022)

Montgomery, David, *Growing a Revolution: Bringing Our Soil Back to Life* (WW Norton, 2017)

Muirhead, M.R., Alexander, T.J.L., Carr, J. (ed.) *Managing Pig Health 2nd Edition: A Reference for the Farm* (5m Books, 2013)

Murray-White, James, 'Reconnecting to a Mythic Mycelium of Place', https://dark-mountain.net/reconnecting-to-a-mythic-mycelium-of-place/

Nuthall, Peter, *The Intuitive Farmer: Inspiring Management Success* (5m Books, 2011)

Organic Research Centre, *Organic Farm Management Handbook 2003* (Organic Farmers & Growers, 2023)

Page, Phillipa and Hamer, Kim, Sheep Keeping (5m Books, 2011)

Peitz, Beate and Peitz, Leopold, *Keeping Chickens 9th Edition: Practical Advice for Beginners* (5m Books, 2016)

Perkins, Richard, *Regenerative Agriculture – A Practical Whole Systems Guide to Making Small Farms Work* (self-published, 2015)

Pretty, Jules, *The Low-Carbon Good Life* (Taylor & Francis, 2022)

Raskin, Ben, *The Woodchip Handbook: A Complete Guide for Farmers, Gardeners and Landscapers* (Chelsea Green, 2021)

Saladino, Dan, *Eating to Extinction: The World's Rarest Foods and Why We Need to Save Them* (Jonathan Cape, 2021)

Salatin, Joel, *Pastured Poultry Profit$* (Chelsea Green, 2013)

Scott, Roger and Oliver, Lee-Anne, *The Veterinary Book for Beef Farmers* (5m Books, 2021)

Sheldrake, Merlin, *Entangled Life: How Fungi Make Our Worlds, Change Our Minds and Shape Our Futures* (The Bodley Head, 2023)

Shiva, Vandana, *Agroecology and Regenerative Agriculture: Sustainable Solutions for Hunger, Poverty, and Climate Change* (Synergetic Press, 2022)

Shorrocks, Victor M., *Conventional and Organic Farming: A Comprehensive Review through the Lens of Agricultural Science* (5m Books 2017)

Shrubsole, Guy, *Who Owns England? How We Lost Our Green and Pleasant Land and How to Take It Back* (William Collins, 2020)

Smaje, Chris, *A Small Farm Future* (Chelsea Green, 2020)

Smaje, Chris, *Saying No to a Farm Free Future* (Chelsea Green, 2023)

Soil Association, *The Agroforestry Handbook: Agroforestry for the UK*, https://www.soilassociation.org/media/19141/the-agroforestry-handbook.pdf

Spector, Tim, *Food for Life: The New Science of Eating Well* (Jonathan Cape, 2022)

Spencer, Kari, *City Farming: A How-to Guide to Growing Crops and Raising Livestock in Urban Spaces* (5m Books, 2017)

Stanley, Joe, *Farm to Fork: The Challenge of Sustainable Farming in 21st Century Britain* (Quiller, 2021)

Starmer, Georgina, Smallholding: *A Beginner's Guide to Raising Livestock and Growing Garden Produce* (5m Books, 2010)

Strand, Sophie, *The Flowering Wand: Rewilding the Sacred Masculine* (Inner Traditions, 2022)

Sustain, Unpicking food prices: Where does your food pound go, and why do farmers get so little?, https://www.sustainweb.org/reports/dec22-unpicking-food-prices/

Tree, Isabella, *The Book of Wilding: A Practical Guide to Rewilding, Big and Small* (Bloomsbury, 2023)

Van Tulleken, Chris, *Ultra-Processed People: Why Do We All Eat Stuff That Isn't Food...and Why Can't We Stop?* (Cornerstone, 2023)

Villalba, Juan (ed.), *Animal Welfare in Extensive Production Systems* (5m Books, 2016)

Winkelmann, Johannes, *Sheep and Goat Diseases 4th Edition: Veterinary Book for Farmers and Smallholders* (5m Books, 2017)

Index

Unbreaking our planet and our systems

Unbreaking is a new imprint from agriculture and veterinary publisher 5m Books focusing on the regenerative action we need to take in response to the varied and interlinked challenges we face, encompassing agroecological, socio-political, economic and financial regeneration.

Unbreaking launches with the publication of
Six Inches of Soil.

Forthcoming titles will include:

Intelligent Farming: Regenerative Change,
How to and Why You Should
Tim Parton

Why Farm at a Loss? Balancing Nature and Energy
through Maximum Sustainable Output
Chris Clark & Brian Scanlon

The Organic Farm Management Handbook 2023 from the Organic Research Centre is the only source of information on the costs and performance of organic farming.

One of the key barriers is the lack of current information on the costs and business performance of organic farms and related management issues that is vital to anyone contemplating the seismic shift to organic farming. The revised OFMH will provide that information – utilising in depth historical data and expert opinion.

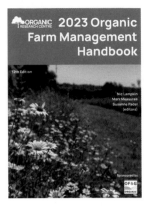

It is the essential tool for understanding the economics of organic farming in the UK. It can:

- Help with business plans and budgets
- Provide a means of assessing the viability of specific crops and livestock
- Advise on conversion related innovations such as new marketing approaches.

Published by the Organic Research Centre, distributed by 5m: available from the 5m Books website, all good bookshops and the ORC's webshop.

With the Land marks 10 years of the Landworkers' Alliance. It explores what it means to work with the land, reflects on the wider land work movement and celebrates what is achievable through collective action.

With the Land is a seed store of stories and poetry, interviews, recipes, essays, artwork and song, by and about people who have tilled and cared for the lands of Britain to produce food, fuel and timber within a culture of regeneration. Old hands and young voices, activists and campaigners, foresters and farmers, shepherds and soil keepers, have come together to create a testament of the collaborative spirit, vision and hard work that goes into restoring our relationships with the natural world, and making a new approach to land use and food growing possible.

It links the past, present and future by bringing together voices from our membership and beyond.

Filled with song lyrics, texts, photography, poetry, a letter and recipe the individual pieces in the book cover issues such as access to land, the importance of seed sovereignty, gender and land work, as well as the ecological and social imperative of creating a better food and land work system by working in harmony with nature.

Spanning landscapes that range from forests to uplands, city farms to rural hinterlands, each section tells the story of how a dedicated network of landworkers are pushing against history to create a radical future-looking movement.

Published by the Landworkers' Alliance, distributed by 5m: available from the 5m Books website, all good bookshops and the Landworkers' Alliance's webshop.